What Your Colleagues Are Saying . . .

A very compelling set of fresh ideas that prepare educators to turn the corner on advocating for social justice in the mathematics classroom. Each book is full of engaging activities, frameworks, and standards that center instruction on community, worldview, and the developmental needs of all students—a much-needed resource to reboot our commitment to the next generation.

—Linda M. Fulmore
TODOS: Mathematics for ALL
Cave Creek, AZ

A wonderful collection of lessons, submitted by teachers, to help students of all ages see topics they care about, and use mathematics as a tool for progress in the world.

—Jo Boaler
The Nominelli-Olivier Professor of Education (Mathematics)
Stanford University, Stanford, CA

Upper Elementary Mathematics Lessons to Explore, Understand, and Respond to Social Injustice is an outstanding addition to the growing number of texts and projects that weave the teaching of mathematics and social justice together. The authors go deep and broad to show how, why, and when this combination of curricular topics improves our students' mathematical understandings while honing their abilities and dispositions to promote social and environmental justice in their own lives and communities.

—Bob Peterson
Editor of *Rethinking Schools*
Editor of *Rethinking Mathematics: Teaching Social Justice by the Numbers*
Milwaukee, WI

Teaching mathematics for social justice affirms the relevance of mathematics instruction to the "real world" and equips educators and students to turn their engagement with mathematics concepts into positive social action. Equal parts approachable and challenging, the lessons get students thinking critically about how mathematics helps them to understand, identify injustice, and develop the skills and confidence to right it.

—Jonathan Tobin
Curriculum and Training Specialist
Learning for Justice, Atlanta, GA

This is the book so many of us in upper elementary mathematics have been waiting for. It's practical, justice oriented, and student-centered. For elementary school teachers looking to integrate social justice lessons with a relevant and timely lens, this book will be instantly applicable to your practice. For everyone else, this book demonstrates that social justice mathematics is critical to the work we must do for our students, our communities, and our profession as mathematics teachers!

—**José Luis Vilson**
Veteran Educator and Executive Director
EduColor, New York, NY

This book is a much-needed and timely resource for teachers, coaches, school leaders, and teacher educators. The authors offer a wide array of lessons that get to the heart of teaching mathematics for social justice for students in Grades 3–5. The diverse topics share a common thread: a commitment to students' learning grounded in meaningful and relevant explorations.

—**Marta Civil**
Professor and Roy F. Graesser Chair
The University of Arizona, Tucson, AZ

I imagine many people will purchase this book for the sample lesson plans. And you should; they're fabulous. But just as fabulous, and equally important, is the framework the authors lay out for a comprehensive, holistic, transformative approach to mathematics teaching, with social justice at its core.

—**Paul C. Gorski**
Lead Equity Specialist
Equity Literacy Institute, Columbia, SC

As a teacher educator for social justice, I am familiar with the near-constant refrain of "this isn't something you can do in math!" This book illustrates just the opposite. Indeed, not only is it possible to engage in social justice mathematics, but it is an educational imperative to do so. This much-needed and valuable collection provides practitioners with clear and compelling lessons that are grounded in theories of justice and equity. Especially timely in this text is the clear evidence that not only can upper elementary–aged children engage in critical conversations, problem solving, and sociocultural analysis in their mathematics classes, but they must. The editors and contributors to this volume have curated a powerful resource that is a must-read for all mathematics educators and those who care about social justice teaching and learning.

—**Alyssa Hadley Dunn**
Associate Professor of Teacher Education
Michigan State University, East Lansing, MI

Upper Elementary Mathematics Lessons to Explore, Understand, and Respond to Social Injustice

at a Glance

In addition to pedagogical tools, additional resources, and voices from the field, this book delivers 15 lessons with extensive additional resources.

Notes tying each lesson back to Social Justice Outcomes, Mathematics Essential Concepts, and Mathematical Practices.

General overview of the lesson describing the background, learning goals, and needed materials and resources to complete the lesson.

SOCIAL JUSTICE OUTCOMES

- I know that the way groups of people are treated today, and the way they have been treated in the past, is a part of what makes them who they are. (Diversity 10)

- I know when people are treated unfairly, and I can give examples of prejudiced words, pictures, and rules. (Justice 12)

- I know about the actions of people and groups who have worked throughout history to bring more justice and fairness to the world. (Justice 15)

- I will speak up or do something when I see unfairness, and I will not let others convince me to go along with injustice. (Action 19)

MATHEMATICS CONCEPTS

- Interpret data distributions to answer questions and pose further questions.

- Use the number line for representing fraction (or decimal) magnitudes and operations.

MATHEMATICS PRACTICES

- Model with mathematics.

LESSON 6.1 "TU LUCHA ES MI LUCHA": MATHEMATICS FOR MOVEMENT BUILDING

Gloria Gallardo and Cathery Yeh

SOCIAL JUSTICE CONNECTION

The United Farm Workers of America (UFW) is the nation's first enduring and largest farmworkers union. This lesson will help students connect the inequities faced by farmworkers then to the inequities in labor rights that continue today. Learning about these struggles through the context of mathematics provides students an opportunity to understand the power in using mathematics to analyze (in)justice and build movements for social justice. There is power in numbers and using mathematics to bridge differences.

DEEP AND RICH MATHEMATICS

This lesson was designed to introduce students to decimal concepts and operations. Students learn about decimal concepts and engage in decimal calculations to examine and compare the differential wages of farmworkers and how numbers were used to build solidarity among the different groups of farmworkers.

Resources and Materials

- Book: *Journey for Justice: The Life of Larry Itliong* by Dawn B. Mabalon and Gayle Romasanta

- Video: "AAPI Civil Rights Heroes—Asian Americans Advancing Justice," from the Zinn Education Project (https://bit.ly/32F9r1N) (Scroll down the page to find the video.)

- Article: "Farmworker Wages in California," by Philip Martin and Daniel Costa, *Economic Policy Institute Working Economics Blog*, March 21, 2017 (https://bit.ly/3rMdS5E)

- Article: "Mapping UFW strikes, boycotts, and farm worker actions 1965–1975," by Katie Anastas, Civil Rights and Labor History Consortium (https://bit.ly/3EBYHiO)

- PBS Documentary: *The Farm Worker Movement* (https://bit.ly/2ZIjzWq)

LESSON 1 FACILITATION

Introduction

Launch Part I (10 minutes)

- In this part of the lesson, the students are introduced to a problem context and invited to consider how the problem is situated in a broader social justice context.

- Orient students to the issue of library resources: *Today we will discuss how books can be distributed to school libraries.* Introduce the following three pictures to students and ask:

  + *What do you notice?*

  + *What questions do you have?*

Source: unsplash.com/@kazuend

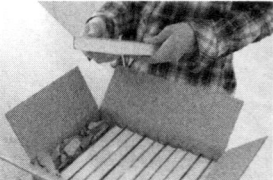
Source: makasana/iStock.com

Source: Yevhen Roshchyn/iStock.com

Explore (20 minutes)

- **Optional:** Have students watch the TED video, "How America's Public Schools Keep Kids in Poverty," by Kandice Sumner (https://bit.ly/3lMxNh4). Discuss the following:

  + *What do you notice?*

Extensive facilitation notes help educators run through the lesson with their class in a thoughtful manner.

Find notes from the authors on communicating with stakeholders to help you and your students share out findings or artifacts with family members and their school community.

COMMUNICATING WITH STAKEHOLDERS

Students may propose their methods to the school administrators and other stakeholders. Students can collaboratively work to share the resource disparities they notice and propose fairer ways to distribute new books (and types of books) to schools.

ONLINE RESOURCES

Available for download at **resources.corwin/TMSJ-UpperElementary**

All worksheets and online resources provided in the Resources and Materials of each lesson are available for download or viewing on the companion website.

▲
Teacher Resource 1: Three Images of Library Books

▲
Teacher Resource 2: Data Tables

ABOUT THE AUTHORS

Find lesson background information and contributor bios at the end of each lesson to give you additional context of how this lesson came to be.

Hyunyi Jung has studied ways to connect mathematics with students' lives through mathematical modeling and culturally sustaining mathematics pedagogy. As a first-generation college student and bilingual learner, she has been interested in the ways in which a broad social system influences mathematical learning and teaching.

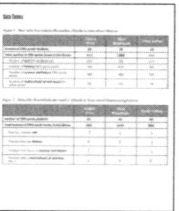
Megan Wickstrom is an associate professor of mathematics education in the Department of Mathematical Sciences at Montana State University where she teaches current and future K–12 teachers. As a former middle school mathematics teacher, she finds joy in working alongside classroom teachers to develop tasks that are relevant and important to students and our communities. She works collaboratively to develop tasks that honor students' lived experiences, support their mathematical identities, and allow them to see the world through different perspectives.

Upper Elementary

MATHEMATICS LESSONS
TO EXPLORE, UNDERSTAND, AND RESPOND TO
Social Injustice

Upper Elementary
MATHEMATICS LESSONS
TO EXPLORE, UNDERSTAND, AND RESPOND TO
Social Injustice

Tonya Gau Bartell · Cathery Yeh · Mathew D. Felton-Koestler
Robert Q. Berry III · and Colleagues

Brian R. Lawler, *Series Editor*

Foreword by Julia M. Aguirre

A JOINT PUBLICATION

For information:

Corwin
A SAGE Company
2455 Teller Road
Thousand Oaks, California 91320
(800) 233–9936
www.corwin.com

SAGE Publications Ltd.
1 Oliver's Yard
55 City Road
London, EC1Y 1SP
United Kingdom

SAGE Publications India Pvt. Ltd.
B 1/I 1 Mohan Cooperative Industrial Area
Mathura Road, New Delhi 110 044
India

SAGE Publications Asia-Pacific Pte. Ltd.
18 Cross Street #10–10/11/12
China Square Central
Singapore 048423

President: Mike Soules
Associate Vice President and Editorial Director:
    Monica Eckman
Publisher: Erin Null
Content Development Editor: Jessica Vidal
Editorial Assistant: Nyle De Leon
Production Editor: Tori Mirsadjadi
Copy Editor: Christina West
Typesetter: Integra
Proofreader: Dennis Webb
Indexer: Integra
Cover Designer: Scott Van Atta
Marketing Manager: Margaret O'Connor

Library of Congress Cataloging-in-Publication Data
Names: Bartell, Tonya Gau, author. | Yeh, Cathery, author. |
    Felton-Koestler, Mathew D., author. | Berry, Robert Quinlyn, III,
    author. | Lawler, Brian (Educator), author.
Title: Upper elementary mathematics lessons to explore, understand, and
    respond to social injustice / Tonya Bartell, Cathery Yeh, Mathew Felton-
    Koestler, Robert Q. Berry III, Brian R. Lawler.
Description: First edition. | Thousand Oaks, California : Corwin, a Sage
    Company, [2023] | Series: Corwin mathematics series
Identifiers: LCCN 2022000382 (print) | LCCN 2022000383 (ebook) |
    ISBN 9781071845516 (paperback acid-free paper) |
    ISBN 9781071883655 (adobe pdf)
Subjects: LCSH: Mathematics--Study and teaching (Elementary school) |
    Social justice and education.
Classification: LCC QA135.6 .B375 2023 (print) | LCC QA135.6 (ebook) |
    DDC 372.7/044--dc23/eng20220711
LC record available at https://lccn.loc.gov/2022000382
LC ebook record available at https://lccn.loc.gov/2022000383

This book is printed on acid-free paper.

22 23 24 25 26 10 9 8 7 6 5 4 3 2 1

CONTENTS

Visit the companion website at
resources.corwin/TMSJ-UpperElementary for
downloadable resources.

Note From the Publisher: The authors have provided video and web content throughout
the book that is available to you through QR (quick response) codes. To read a QR code,
you must have a smartphone or tablet with a camera. We recommend that you download
a QR code reader app that is made specifically for your phone or tablet brand.

FOREWORD

Welcome to a new generation of social justice mathematics advocates.

I am very excited and humbled to be writing the Foreword to this series of books about teaching mathematics for social justice. Over the past 30 years, there have been very few teaching resources available in one place that support teachers to embrace children as mathematical problem posers, sensemakers, and community change agents. The volumes in this series provide teachers with pedagogical and curricular tools to create mathematical learning environments that invite curiosity, social consciousness and critique, and mathematical analysis and innovation—multiple paths that lead to challenging societal inequities and making mathematics a more humanizing and just experience.

Why have such resources taken so long? This is an excellent and complex question. In my over 25 years of experience as a mathematics educator and scholar with an explicit equity ethic (McGee, 2020), the mathematics education community has been slow to embrace an equity and justice-centered approach to mathematics education. There has been a tremendous amount of emphasis on reform mathematics, mathematical thinking, and mathematical discourse. Yet that same approach has reinforced beliefs and structures about mathematics being universal and culture free. It was about making dominant mathematics—which emphasizes cisgendered male-centric and euro-centric values—accessible to more people, while failing to acknowledge that mathematics has been created and communicated by cultures and communities across the globe since time immemorial. Efforts to introduce mathematical investigations that center on community, family, and ancestral knowledge and uses of mathematics into schools—especially activities that mathematized fairness, representation, and power relationships—brought consistent criticism from mathematicians and mathematics educators, who asked, "Where's the math?" Ironically, it is the powerful mathematics inherent within these investigations that brought to light injustice, inequity, and a demand for change to make things right. Mathematics has always been created by us, with us, and for us. It is time that we embrace this idea fully.

The link between social justice and mathematics has strong roots in liberatory education as practiced by Brazilian educator and philosopher Paolo Freire, and connected to mathematics by American mathematics education scholars Marilyn Frankenstein, Arthur Powell, Eric Gutstein, Rochelle Gutiérrez, and Danny Martin. However, for me, another crucial link made between mathematics and social justice was the work of civil rights leader Bob Moses and the work of the Algebra Project. Bob Moses passed away in the summer of 2021 at the age of 86. He was the first mathematics educator to help me see mathematics as a civil rights issue. In his book *Radical Equations* (Moses & Cobb, 2001), he argued that "full citizenship" in the 21st century—including economic and political access as well as informed decision making and civic engagement—in our society is inextricably linked to mathematics literacy. We must remove the systemic barriers, beliefs, and structures that deny children the right to have a high-quality, nonviolent, and meaningful mathematics education. The guidance and resources found in this series support teachers to do just that.

I would be remiss if I did not acknowledge that this series was developed during the twin pandemics: COVID-19 and systemic racism. COVID only exacerbated systemic racism inherent in the education, economic, political, health care, and legal systems of the United States. Disproportionate deaths due to COVID in communities of Color; the deaths of Breonna Taylor and George Floyd, among countless other Black and Brown people at the hands of the state; the rise of hate crimes, domestic violence, and sexual assault; increased gun violence and homelessness; the opioid crisis; the separation of families at the border; the rise of suicide among young people and veterans; anti-Asian and anti-immigrant hate; and missing and murdered Indigenous women are some examples of the pain and violence we have endured. Our planet is on fire, and many of us lack access to clean water, air, and earth. We must change and we must look to the next generation of young people to lead this effort.

This series continues the work of social justice mathematics advocacy by providing classroom-based mathematics lessons that build children's empathy and analysis skills to connect mathematics to their own lives, their communities, and the complex world around them. Relationships among people, animals, and the planet are mathematized in various ways. The investigations are grounded in social justice and mathematics standards so educators can be confident that the work meets multiple teaching goals. But what I am really excited about is the amplifying of youth voice and activism through these mathematics activities. Bob Moses said,

> We don't listen to kids enough. Really listen. It is a difficult thing for grown-ups to do—listen and actually pay attention to what young people are saying. In the Algebra Project we are still learning how to do this also. It is the voices of young people I hear every day, more than anything, that gives me hope. (Moses & Cobb, 2001, p. 191)

We must listen and learn from the voices of young people. Children can grapple with hard topics because they understand ideas of fairness, sharing, love, and friendship. They are curious about the world and they reflect the world. They are complex human beings. They are our future. They are our hope. Welcome to the new generation of social justice mathematics advocates.

—Julia M. Aguirre, PhD
December 3, 2021

ACKNOWLEDGMENTS

First, we—Tonya, Cathery, Matt, and Robert—would like to thank all of the lesson authors for your willingness to write and share lessons so that this book would become a reality. We knew from the start the importance of having a diverse set of voices from the field—educators who could share lessons and experiences, and who were willing to trust us with their work. Thank you.

Second, thank you to all of the educators who field-tested or reviewed a social justice mathematics lesson (SJML). Your feedback and suggestions helped further develop key features for the lessons included in this book. Your willingness to share your experiences with these SJMLs and other social justice lessons has helped enhance our thinking as we developed the chapters in this book.

Next, we would like to thank all of the reviewers. Your thoughtful and thorough feedback on the chapters and lessons provided us with a road map to make the much-needed revisions. When you read the final version of this book, our hope is that you see we have attempted to address your concerns and suggestions.

Thank you to Corwin and NCTM for your willingness to publish this book.

We thank Erin Null, Jessica Vidal, and the rest of the Corwin team. Beyond feedback, you provided us with encouragement and guidance. Thank you to your team for your support in publishing a book that seeks to equip elementary teachers of mathematics so that they can equip their students with the mathematical tools to explore, understand, and respond to social injustices as they become powerful learners and doers of mathematics.

Last but not least, we would like to thank the Series Editor, Brian R. Lawler, for his endless encouragement, insight, nudging, and leadership. You kept us going when we didn't think we could.

To everyone above, a special thank you as all of this happened amid the global COVID-19 pandemic, the current fight for Black, Brown, Indigenous, and Asian lives, and the rise in anti-CRT laws. Your communities, classrooms, and lives were upended; yet these challenges and your commitment to students, their learning, and social justice gave us strength and determination to contribute to this process.

Tonya's Special Acknowledgments

To mathematics teachers, educators, colleagues, and friends in my career—and especially to Cathery, Matt, and Robert—I have learned so much from you and am a better person and educator for it.

A special thank you to Val, who reminds me daily of the important work teachers do, and to Jaedyn and Rich for their love, support, and enthusiasm for my work and for joining me on this journey to make mathematics teaching and learning meaningful and in support of social justice.

Cathery's Special Acknowledgments

Tonya, Matt, and Robert, I have learned immensely from writing with you. Thank you for modeling social justice work as a process that is lived, not just experienced in text.

Writing has become an act of liberation, a sharing of learnings from the teachers, students, and communities around me and the elders that came before me. I hope this book does justice to your loving and visionary work! To Em, El, and David, thank you for sitting with me to keep me company as I write and for reminding me daily that our words matter. Tú eres mi otra yo.

Matt's Special Acknowledgments

It has been deeply rewarding to collaborate with such thoughtful and talented colleagues. Tonya, Cathery, and Robert, you have pushed my own thinking about equity and justice both in and outside of the field of mathematics education. Thank you to my family and friends who also challenge my ideas and often serve as a sounding board for my thinking.

Robert's Special Acknowledgments

Thank you to all the teachers and mathematics educators who inquired, advocated, and supported the vision for this book. Tonya, Matt, and Cathery, you all have been truly wonderful and have made me a better mathematics educator.

Finally, we extend a very special acknowledgment to Brian, Basil, John, and Robert for the vision and work from their book, *High School Mathematics Lessons to Explore, Understand, and Respond to Social Injustice* (Berry et. al, 2020), which is the inspiration for this and the other books in this series.

PUBLISHER'S ACKNOWLEDGMENTS

Corwin gratefully acknowledges the contributions of the following reviewers:

Kyndall Allen Brown
Executive Director
California Mathematics Project, Los Angeles, CA

Theodore Chao
Associate Professor
The Ohio State University, Dublin, OH

Jenna Laib
K–8 Mathematics Specialist
Brookline Public Schools, Medford, MA

ABOUT THE AUTHORS

Tonya Gau Bartell: I am currently an associate professor of mathematics education in the College of Education at Michigan State University and serve as the Associate Director of Elementary Programs. I started teaching 25 years ago, beginning my work as a high school mathematics teacher for 6 years, including 3 years as a founding teacher in an alternative high school to support students labeled as not succeeding by the system. To this day, the students I taught in that setting remain some of the brightest and most creative individuals I have had the pleasure to know. For the last 15 years, I have volunteered in elementary mathematics classrooms and studied elementary mathematics education. Throughout my career, I have facilitated professional development with teachers and been an instructor of prospective mathematics teachers as they aim to explicitly consider equity and justice in order to meet the needs of all learners in inquiry-based, heterogeneous, equitable classrooms.

I am passionate about learning about and supporting teachers in developing equitable mathematics instructional practices that recognize and transgress systemic inequity. I understand that issues of culture, race, ethnicity, identity, and power influence students' opportunities to learn and teachers' opportunities to teach mathematics and that these factors must be explicitly discussed and addressed if we hope to fully support equitable mathematics teaching and learning. I am honored to have participated in the writing of this book and in continued efforts supporting mathematics education that explores, understands, and responds to social injustice and supports students' learning of mathematics.

Cathery Yeh: My students, their parents/caregivers, and the communities I have had the privilege to work with remind me daily that students' identities and their sense of belonging shape learning. I started teaching 24 years ago, beginning my tenure in dual-language classrooms in Los Angeles and abroad in China, Chile, Peru, and Costa Rica. As a classroom teacher, I made home visits to every student home (over 300) and co-taught mathematics lessons with parents/caregivers and community organizers to integrate students' lived experiences, knowledge, and identities into the curriculum. As a learner of mathematics, my own schooling mirrors my research commitments to bilingualism, culturally sustaining pedagogies, and ethnic studies. I came to the United States at the age of 5. I was the only emergent bilingual student in class. At the end of the school year, I was retained. Kindergarten in California is optional, but I had to take it twice because I was not yet fluent in English. My first year of schooling in the United States highlights how too many of our children feel invisible in the classroom.

As a mathematics education scholar committed to equity and social justice, I acknowledge the many privileges I have and my role as both the oppressor and the oppressed. As a Chinese American scholar, educator, and community organizer living at this very time, I feel the rise of Asian hate crimes, prejudice, xenophobia, and discrimination. The Atlanta shooting is only a recent one in a legacy of anti-Asian violence in the Americas. Orientalist stereotypes of submissiveness, passiveness, and the exoticization of Asian women have not only led to the longstanding history of hypersexualization and violence, but they also have forced false obedience and compliance within the Asian American community. I recognize that oppression does not play out uniformly, but rather across the multitude of intersections of social identities, and it is often disguised through division and intentional pitting of one group against the other (e.g., using Asian Americans to perpetuate the myth of meritocracy and as a wedge against our siblings of Color). I call on both my own Asian American siblings and the mathematics education community to disrupt the silence to injustice. We need to be loud.

What happens outside of the classroom can no longer be ignored. We're in a critical moment where children are seeing injustice but not necessarily understanding the root and stem of the injustice. As educators, we have an obligation to teach and learn with children about these critical and complex issues. My hope is that this book can help spark conversations that attend to identities, histories, communities, and possibilities!

Mathew D. Felton-Koestler: I am currently an associate professor in the Department of Teacher Education at Ohio University in Athens, Ohio, where I primarily teach mathematics methods courses for future elementary and middle school teachers. Before coming to Ohio University, I was in the Department of Mathematics at The University of Arizona where I taught mathematics content courses for future teachers. I received my B.S., M.A., and PhD from the University of Wisconsin–Madison.

I began my focus on education working with my former elementary teacher, Mazie Jenkins, to engage elementary students in exploring geometry through quilt design. Throughout my career I have benefited from opportunities to collaborate with practicing teachers in the classroom and in professional development settings. I attempt to approach my work with teachers as collaborative and have greatly valued this aspect of my work. I particularly enjoy the challenge of blending rich mathematics with explorations of our social and political world in tasks that are accessible to a broad audience. Prior to COVID-19, I co-ran a summer camp for middle school students to use mathematics to explore social and political issues related to their interests and look forward to returning to similar work in the future.

Robert Q. Berry III: I am the Dean of the College of Education at the University of Arizona and the Paul L. Lindsey & Kathy J. Alexander Chair. I served as President of the National Council of Teachers of Mathematics (NCTM) from 2018–2020. I hold a BS in middle grades education from Old Dominion University, an MAT in mathematics education from Christopher Newport University, and a PhD in mathematics education from the University of North Carolina at Chapel Hill. I taught in public schools and served as a mathematics specialist.

Equity issues in mathematics education are central to my research efforts with four related areas: (a) understanding Black children's mathematics experiences, (b) measuring standards-based mathematics teaching practices, (c) unpacking equitable mathematics teaching and learning, and (d) exploring interactions between technology and mathematics education. I co-edited the 2020 book, *High School Mathematics Lessons to Explore, Understand, and Respond to Social Injustice.* My articles have appeared in the *Journal for Research in Mathematics Education,* the *Journal of Teacher Education, Educational Studies in Mathematics,* and the *American Educational Research Journal.* I am a two-time recipient of NCTM's Linking Research and Practice Publication Award.

INTRODUCTION

Elementary students bring curiosity and wonder to their learning environments, including connections they build between school and their lives outside of school, the wonderings about their world they bring into conversation, and their sense and inclination for problem solving and exploration. Elementary mathematics should be related to students' questions, interests, and lives, and build on their experiences in school, at home, and in their communities. Mathematical investigations should occur in contexts that are interesting and meaningful to students. Teachers' desire to build on this knowledge and expertise in lessons can be stifled based on calls to use prepackaged materials, rigid pacing guides like factory workers on an assembly line, and limited time to plan and collaborate around integrated lessons or units. Yet teachers push forward, mindful of what they know to be best for the groups of students in their care.

We know that when teachers can draw from the lives and interests of students, families, and communities, meaningful instruction happens. This instruction is rooted in contexts that are familiar, personal, and connected to the things that students in the classroom are experts in. This is an idea we share with teachers we collaborate with, preservice teachers we mentor, and policy makers we aim to inform. This work is complex; it takes time, persistence, and creativity. But it is also worthwhile and can mean the difference for students developing a passion for mathematics.

By the time you finish reading and implementing ideas from this book, it is our hope that you will experience several benefits:

- You will acknowledge all children in your classroom as capable mathematics learners and social justice advocates, seeing their experiences and wonderings as opportunities to contribute their voice and agency with their communities.

- You will recognize the ways in which you are already on this journey of teaching mathematics for social justice in your classroom and in your professional development and identify ways to enhance what you are already doing.

- You will learn to use mathematics regularly and seamlessly as a lens to facilitate discourse around important topics that matter to your students' lives.

- Your students will see how mathematics applies to their lives, helps them understand the nature and roots of social disparities, and how they can use it to intervene and seek equity.

- Your students will become more engaged in their communities, cities, states, and regions, and feel empowered to lead efforts to improve social disparities.

- You will know that you are not alone in this work and that there are other elementary mathematics educators who exist to partner with, to share ideas with, or to lend support.

This book offers a collection of multiday mathematics lessons. Understanding mathematics and understanding the world through mathematics takes time. Each lesson is tied to the mathematics standards we must teach as well as to the Social Justice Standards from Learning for Justice (formerly known as the Teaching Tolerance Standards from the Southern Poverty Law Center) and is also grounded in issues of social importance to both you and your students (Learning for Justice, 2016). These lessons are bookended by practical advice. In the opening chapters, we discuss our ideas of what it means to teach mathematics for social justice and strategies to effectively do so. We close by offering some ideas for how to create your own social justice mathematics lessons as well as with some wisdom and advice from other teachers who have embarked on this journey.

WHY IS TEACHING MATHEMATICS FOR SOCIAL JUSTICE CRITICAL?

An important aspect of our responsibility as educators is to help empower students to be agents of change in their communities, states, nations, and world.

While it is not always discussed in classrooms, students regularly experience the effects of social privilege, power, and activism. Each day, students in schools and communities are faced with disparities in opportunity. Parents, caregivers, grandparents, aunts, uncles, friends, cousins, neighbors, and community members share stories about their lives and perspectives and students carry this knowledge inside of themselves. Students have concerns about their world, their community, and their family. An important aspect of our responsibility as educators is to help empower students to be agents of change in their communities, states, nations, and world.

We would like to go further than simply stating the importance of connecting mathematics teaching and learning to teachers' and students' lived experiences and interests; we argue that teaching mathematics for social justice (TMSJ) is critical for four reasons:

- It builds an informed society.

- It connects mathematics with students' cultural and community histories.

- It empowers students to confront and solve real-world challenges that they face.

- It helps students learn to use mathematics as a tool for social change.

TMSJ can and should extend mathematics beyond the classroom. It can and should encourage students to

- Learn important mathematics;

- Express self-love, pride, confidence, and healthy self-esteem about themselves as mathematical thinkers and learners;

- Express comfort in working with and learning from people who are both similar to and different from them and engage respectfully in collaborative work;

- Develop concern for the happiness of other human beings and life forms; and

- Plan and carry out collective action using mathematics as a tool to address injustice in the world.

THIS BOOK'S AUTHORSHIP

This book began when we—the four editors—reacted to a post online announcing the publication of the book, *High School Mathematics Lessons to Explore, Understand, and Respond to Social Injustice*. We were thrilled to see such a resource published, and we almost immediately began commenting on how we wished there were similar resources for students at other grade levels. Others joined in, noting specifically how great it would be to have such a resource for early childhood and early elementary grades, upper elementary grades, and middle school grades.

We already had a mutual interest in providing elementary mathematics teachers with a collection of lessons that address both essential mathematical concepts and issues of social concerns and injustices. We all worked with and heard from many elementary teachers who recognized that students were significantly more engaged when the context of their learning was more personally meaningful. Furthermore, elementary teachers shared how their students regularly posed questions about their own lived experiences as well as injustices that they heard about or experienced. More and more of the teachers with whom we work are using social, economic, and environmental justice contexts in their elementary classrooms, including when teaching mathematics.

Lesson Authors

As we discussed these shared experiences of elementary teachers wanting a similar resource, we recognized that we were all working in universities. Our perspectives represented only a sliver of the teachers and students interested in the book we imagined. Knowing that our own four sets of passions, perspectives, and lived experiences were limited and wanting to include as diverse a range of perspectives and voices as we could manage, we solicited lessons from mathematics educators around the country, with specific requests for lessons created by or with practicing elementary mathematics teachers. In addition, we sent the lessons out to field testers to implement in their own elementary classrooms as well as to reviewers with elementary mathematics teaching experience. These field testers and reviewers gave extensive and valuable feedback on the lessons; their thoughtful insight made these lessons, and this book, better.

We value the voice of each educator who contributed a lesson, and we have made all attempts to share their work and their voice with you. Outside the major required elements for submissions, we asked lesson authors to format and submit lessons based on how they had implemented them in their classrooms. We then edited lessons for clarity, mathematical rigor appropriate for the identified grade level, explicit social justice goals, and cohesion in order to highlight the voice and authenticity of work in the field. Lessons have been tested and refined, in their own classrooms and in others. We are grateful to the lesson authors and all those who helped develop this book. As a team, we each have our own motivations and understandings of TMSJ. Each lesson author is highlighted in the chapter in which you find their lesson, providing readers some information about each of them, including their journey to becoming social justice educators.

WHO IS THIS BOOK FOR?

This book welcomes those who are new to TMSJ and those who have been engaged in this work for some time. This book is meant for Grade 3–5 teachers, mathematics coaches, school and district leaders, and mathematics teacher educators to empower students and teachers alike. We intend for this book to support you and your students to move from questions like "When/How/Where am I ever going to use this?" to questions like "What can we do about this?" "How can mathematics be used as a tool to address this injustice?" Those outside of the Grades 3–5 band may find the lessons and guiding principles as important in their thinking about what experiences students may come to their classrooms with, or as an eye-opening highlight of the capabilities that upper elementary students bring with them to the classroom setting. Administrators may find reflections and lessons learned powerful as ways to conceptualize support for teachers who are on this journey and needing a supportive community (whether on their teaching team, or when facing stakeholder pushback).

When children learn that mathematics can be used as a tool to help them understand, explore, and investigate social situations, they are empowered to see themselves as both mathematical thinkers and active change agents in a world of change.

During your reading, we hope that you will grow in your understanding that mathematics may be a privileged space through which both you and your students can be empowered. Many students do not have an opportunity to connect mathematics with their culture and lives. Thus, your interest, reading of this book, and implementation of lessons in the classroom present an opportunity to shape students' lives and actions. When children learn that mathematics can be used as a tool to help them understand, explore, and investigate social situations, they are empowered to see themselves as both mathematical thinkers and active change agents in a world of change. We hope that the lessons and the critical call for action contained in this book highlight how each and every student is capable of mathematical learning and can be empowered to use mathematics for change in their own and others' lives.

THE BOOK'S ORGANIZATION

This book is organized into three parts. Part I, consisting of Chapters 1–4, provides a foundation for TMSJ. Chapter 1 shares our idea of what social justice means and what TMSJ looks like, framed around the four domains of the Social Justice Standards, laying an important context for understanding the goals of TMSJ and implementing the lessons of this book. We hope it can also serve as a chapter for you to reflect on your own goals for teaching children through mathematics. Chapter 2 begins a conversation about preparing to teach for social justice and foster a classroom community for social justice in your mathematics classroom. This chapter considers mathematics content as well as the local context in which you will implement TMSJ. It also introduces questions of *who*, *how*, and *when*. Chapter 3 aims to support you in leveraging the tools you are already using to implement TMSJ, particularly around goal setting, using inquiry-based/problem-based learning, and formative assessment. Of note, we provide suggestions in this chapter on how to tread across potentially controversial topics and lead difficult discussions. Chapter 4 focuses on teaching social justice mathematics lessons, including discussion of the structure of lessons in this book as well as a pedagogy that engages students in actively investigating, understanding, and reflecting on challenging mathematical and social questions to empower themselves into action.

Part II of the book includes social justice mathematics lessons, organized around three overarching themes: Building and Examining Identities (Chapter 5), Society and Social Movements (Chapter 6), and Understanding Our World (Chapter 7). Each lesson includes reference to a mathematics standard and a mathematics practice to support teachers in easily locating ideas that may be infused with mathematical course progressions from their state, district, or school. Each lesson also references one or more domains from the Social Justice Standards to support teachers in attending to a variety of objectives identified in these standards and to find lessons that tackle certain social injustices that are relative to their demographic, environmental, or social contexts. You may have expected that the book would be organized by grade level or mathematics content. We chose not to organize by grade level because most of the lessons can be adapted to target learning goals for Grades 3, 4, or 5. We chose not to organize by mathematics content because not all of the lessons fit nicely into a content domain. Importantly, we see this as a strength of the lessons—that many support student learning across multiple content domains. Yet we knew that some organization would be helpful for you, the reader. We hope that the overarching themes serve that purpose.

Part III offers two concluding chapters. First, Chapter 8 shares the reflections of some of the contributing authors, including their experiences implementing these lessons and more generally about experiences of TMSJ. And finally, in Chapter 9, we share our recommendations for developing your own social justice mathematics lessons.

If you are like us, you might not read this book straight through from cover to cover. It might make sense to identify the lessons that align with the content standards assigned to the grade you are teaching. Or you might read the introductory chapters (Chapters 1–4), but then skip around Chapters 5, 6, or 7 for lessons and topics that speak to you and may be most relevant to the students in your classroom. While we provide these lessons as templates to use in your classroom, as with any curricular resource, we assume that you will have to think through them and make minor (and maybe major) changes to meet the needs, interests, and backgrounds of the students in your classroom.

We hope that the resources in this book will help you create and focus energy on authentic experiences for students while also generating mathematical analysis or modeling to explore and take action upon issues of injustice relative to students' lives. We commend you for bringing your students' curiosities and concerns about their lives into your mathematics classroom. We hope that the lessons in this book help you to foster student-to-student interactions that move beyond the mathematics to be learned and into actionable change in students' lives and society.

PART

I

TEACHING MATHEMATICS FOR SOCIAL JUSTICE

WHAT IS SOCIAL JUSTICE, AND WHY DOES IT MATTER IN TEACHING MATHEMATICS?

It's early October. Welcome to Ms. Bailey's classroom. Ms. Bailey's students are racially diverse, and the student body is largely middle class. As the fourth graders arrive and get settled in groups of four, Ms. Bailey checks in with them. She asks how a family's gardening project is going and discusses another student's celebration of Eid Milad un-Nabi, an Islamic holiday commemorating the birthday of the prophet Muhammad. Ms. Bailey finds time to connect personally with a few students each day and has spoken with many of their families, so she knows about their lives and concerns.

On the board in both English and Spanish is a prompt: What's Happening in our Community and the World? *(Ms. Bailey invites students to repeat the prompt in their first language, allowing multilingual students more agency and providing the opportunity for all students to learn more languages.)*

As students share ideas, Ms. Bailey probes their thinking by asking why they are interested in certain topics and if anyone wants to share their own experiences with or feelings about a topic. The class generates the following ideas:

> *Plastic in the oceans*
> *Getting on the internet for school*
> *Bus route/travel time to get to school*
> *Not enough to do in our town*
> *Black Lives Matter*
> *Coronavirus*
> *Global warming*
> *Transgender access to restrooms*
> *Pollution*

To ensure that students have multiple modes of participation, Ms. Bailey often uses small groups and pairs in addition to whole-class discussions:

> *Okay, we've got a great list here. Go to your groups. Your job is to talk about this list. Each person should share one or two ideas that they are most interested in or concerned about and why. When you are done, take two sticky notes and vote for your top-two topics.*

Again, the directions are posted on chart paper.

After students finish their group discussions, Ms. Bailey says, "Thank you, everyone. We have some important ideas. I'll use these to help plan some lessons we can do later in the year." One of the ideas with many sticky notes is getting on the internet for school.

* * *

It's now the end of October. Ms. Bailey has been collecting articles and data to develop a lesson exploring access to the internet. In particular, she commits to a stance of considering how internet access is inequitably distributed and how that may affect the most marginalized members of our society. Ms. Bailey has shared some of these articles in class as part of social studies and language arts.

Ms. Bailey is ready to introduce the lesson: "Alright, let's get started. One of the topics you were interested in was how to make sure everyone could use the internet when they need to." The lesson begins with a discussion about what students already know about internet access and problems they or people they know have had with getting access to the internet. Students have opportunities to share about why the internet is or would be important in their families and any rules they have about how they are allowed to use it.

The class reads about internet access and how it differs across the United States as well as in comparison to other countries. They learn about how the internet works and the infrastructure needed for everyone to have access. Students interview family members and friends to learn about whether internet access is important to them and why.

The class creates a survey to learn about the community's internet needs and figure out a plan for how to administer it. They also look at local internet plans that are available in their area and learn about the range of incomes in the area. They use mathematics to create several budget scenarios and to create examples of what families might need to give up in order to afford the internet. A central topic for discussion is the difference between wants and needs and how we should think about internet access in today's world.

The class begins to brainstorm possible solutions, such as working with nonprofit organizations or petitioning the local government to invest in broadband access. They come up with a rough estimate for what a solution might cost. They compare the estimate to other government projects and to how much local taxes would have to go up to pay for it. Finally, groups create presentations to share with families and community organizations. They must represent data they have found from their survey and in articles, costs for local families and whether they are reasonable, and a summary (with mathematical information) of a proposed solution.

Ms. Bailey's classroom has several features that are important in teaching mathematics for social justice (TMSJ). She provides students with multiple opportunities to engage with the content and form strong connections with students, families, and the community. She builds on what students know and anticipates what may come up in activities. She provides opportunities to affirm students' experiences and identities, value a diverse range of perspectives, and reflect on issues of (in)justice. She provides opportunities to take action. The lessons in this book provide a starting place for engaging your own students in similar ways. As we discuss throughout the book, you will need to consider your students, families, and context in deciding how to introduce and adapt these ideas.

WHAT DO WE MEAN BY SOCIAL JUSTICE?

Social justice demands equity for all people while acknowledging the variety of experiences, values, and worldviews that exist from diverse perspectives. Social justice means committing to challenging the social, cultural, and economic inequalities that arise from the differential distribution of power, privilege, and resources in our world. Social justice instills personal responsibility to collaborate with others for the common good and to improve institutions so they can support personal and social development. The United Nations (2006) contends that social justice must start with all human beings' right to benefit from a safe and pleasant environment, which entails the fair distribution of the fruits of economic growth. Jost and Kay (2010) provided a general definition of social justice as the conditions in which

1. Benefits and burdens in society are distributed per some set allocation principles;

2. Procedures, norms, and rules that govern decision-making preserve the basic rights, liberties, and entitlements of individuals and groups; and

3. Human beings are treated with dignity and respect by authorities and others.

Social justice emphasizes the fair and just relations between the individual and society, meaning that society has a responsibility to ensure equal rights, equal opportunities, and equal treatment for each and every individual.

These definitions of social justice require us to think about how people connect, how resources are distributed, and the meaning of fairness. For us, social justice emphasizes the fair and just relations between the individual and society, meaning that society has a responsibility to ensure equal rights, equal opportunities, and equal treatment for each and every individual. When relations are not equal, we must act and respond to instances of injustice.

In this book, we draw on the Social Justice Standards (formerly known as the Teaching Tolerance Standards from the Southern Poverty Law Center) to inform our understanding of social justice (Learning for Justice, 2016). The standards include anchor standards and learning outcomes grouped by grade band (K–2, 3–5, 6–8, and 9–12) and divided into the four domains of Identity, Diversity, Justice, and Action. They provide a common language and organizational structure to guide curriculum development and implementation to prepare students to actively participate in a diverse democracy (see Figure 1.1).

Figure 1.1. Anchor Standards and Domains

Anchor Standards and Domains

IDENTITY

1. Students will develop positive social identities based on their membership in multiple groups in society.
2. Students will develop language and historical and cultural knowledge that affirm and accurately describe their membership in multiple identity groups.
3. Students will recognize that people's multiple identities interact and create unique and complex individuals.
4. Students will express pride, confidence and healthy self-esteem without denying the value and dignity of other people.
5. Students will recognize traits of the dominant culture, their home culture and other cultures and understand how they negotiate their own identity in multiple spaces.

DIVERSITY

6. Students will express comfort with people who are both similar to and different from them and engage respectfully with all people.
7. Students will develop language and knowledge to accurately and respectfully describe how people (including themselves) are both similar to and different from each other and others in their identity groups.
8. Students will respectfully express curiosity about the history and lived experiences of others and will exchange ideas and beliefs in an open-minded way.
9. Students will respond to diversity by building empathy, respect, understanding and connection.
10. Students will examine diversity in social, cultural, political and historical contexts rather than in ways that are superficial or oversimplified.

JUSTICE

11. Students will recognize stereotypes and relate to people as individuals rather than representatives of groups.
12. Students will recognize unfairness on the individual level (e.g., biased speech) and injustice at the institutional or systemic level (e.g., discrimination).
13. Students will analyze the harmful impact of bias and injustice on the world, historically and today.
14. Students will recognize that power and privilege influence relationships on interpersonal, intergroup and institutional levels and consider how they have been affected by those dynamics.
15. Students will identify figures, groups, events and a variety of strategies and philosophies relevant to the history of social justice around the world.

ACTION

16. Students will express empathy when people are excluded or mistreated because of their identities and concern when they themselves experience bias.
17. Students will recognize their own responsibility to stand up to exclusion, prejudice and injustice.
18. Students will speak up with courage and respect when they or someone else has been hurt or wronged by bias.
19. Students will make principled decisions about when and how to take a stand against bias and injustice in their everyday lives and will do so despite negative peer or group pressure.
20. Students will plan and carry out collective action against bias and injustice in the world and will evaluate what strategies are most effective.

Source: Reprinted with permission of Learning for Justice, a project of the Southern Poverty Law Center. www.learningforjustice.org.

Teaching about Identity, Diversity, Justice, and Action supports teachers in engaging in a range of anti-bias and social justice issues. The Anti-Defamation League (2021) defines anti-bias education as

> *an approach to teaching and learning designed to increase understanding of differences and their value to a respectful and civil society and to actively challenge bias, stereotyping and all forms of discrimination in schools and communities. It incorporates an inclusive curriculum that reflects diverse experiences and perspectives, instructional methods that advance all students' learning, and strategies to create and sustain safe, inclusive, and respectful learning communities.* (para. 1)

Two anti-bias concepts underpin the Social Justice Standards: prejudice reduction and collective action. Prejudice reduction refers to a collection of techniques to break down destructive stereotypes. Collective action refers to the coordinated work that traditionally marginalized and oppressed groups do together to demand justice and equity.

The emphasis on an actively anti-biased curriculum resonates with the joint position paper by NCSM and TODOS: Mathematics for ALL (2016) as well as the position paper of the Benjamin Banneker Association (2017). Both papers argue that embracing social justice requires moving beyond just *noticing* injustices to include actions that confront oppression and marginalization—hence, the title of the book: *Lessons to Explore, Understand, and Respond to Social Injustice*.

PAUSE AND REFLECT

What are some guiding principles that underpin your teaching philosophy?

WHAT IS TEACHING MATHEMATICS FOR SOCIAL JUSTICE?

To illustrate our conception of TMSJ, in this section we draw from the four domains of the Social Justice Standards—Identity, Diversity, Justice, and Action—in the context of mathematics (see Figures 1.2, 1.3, 1.4, and 1.5). Let's look at both teacher and student roles in each domain, drawing on the work of Dingle and Yeh (2021), Learning for Justice (2021), and Yeh (under review).

Domain 1: Identity

In the Identity domain, the teacher provides students with opportunities to learn about who they are and where they are from. This set of Social Justice Standards seeks to reduce bias as students learn about their own identities, privileges, and responsibilities. The goal is for students to develop positive social identities and to understand that multiple identities intersect to create unique and complex individuals. Students' sense of self is integral to their learning of mathematics.

As Banks (2004) argues, marginalized communities often have powerful forms of knowledge and thinking that challenge existing political, economic, and educational practices. Yet many of these rich and collective ways of knowing and being have not yet been made visible in mathematics education. Figure 1.2 describes teacher and student roles that reflect the Identity domain in the context of mathematics teaching and learning.

Figure 1.2. Identity Domain in the Context of Mathematics Teaching and Learning

IDENTITY
Teachers:
• Recenter identities, perspectives, and knowledge traditions that have often been silenced.
• Attend to and honor students' multiple social identities in curricular design and its implementation.
• View students as competent mathematical beings whose lived experiences and community and cultural ways of knowing are leveraged during mathematics instruction.
• Deconstruct negative stereotypes about students' mathematical identities and about who can and cannot do mathematics.
Students:
• Recognize that people's multiple identities interact and create unique and complex individuals that contribute to their learning of mathematics.
• Develop language and historical and cultural knowledge to affirm and describe their membership in multiple identity groups and their contributions to mathematics.
• Express self-love, pride, confidence, and healthy self-esteem about themselves and their community as mathematical thinkers and learners.
• Recognize the traits of the dominant culture, their own culture and other cultures, and how to negotiate their own identity.

Source: Adapted from Learning for Justice (2021).

 Available for download at **resources.corwin/TMSJ-UpperElementary**

Teachers can engage students in the process of honoring students' identities in mathematics by drawing on genuine problems from students' everyday lives. For example, Mrs. Mayra Orozco and her fifth-grade team begin each school year by asking students to video record a short interview in which they ask a family member in English and their home language (if multilingual), "Tell me about yourself. How do you use mathematics every day?" These brief clips are used to create mathematics lessons that honor students' identities and their lived experiences (Yeh et al., 2017).

The mathematics classroom should support students in cultivating a sense of pride in their culture, heritage, race and ethnicity, religion, skin tone, gender, and other group identities. Students should develop language and historical and cultural knowledge about different aspects of their identity, the identities of their peers, and the history associated with it. Liz Esqueda, a third-grade teacher in Oakland, California, develops students' understanding of place value through a lesson in which her students learn about how numbers were represented on Quipus as well as in comparison to the Hindu-Arabic number system. The lesson allows students to explore the mathematics rooted in ancient histories of people and empires of Color[1] while deepening their understanding of our base-10 number system structure. When students are supported to learn more about their own history and others', they are better able to identify, deconstruct, and challenge harmful stereotypes about themselves and others.

Engaging and valuing identity requires teachers to also recognize that inequities in the larger society are replicated in structures and practices that perpetuate disparities in learning opportunities based on race, class, language, gender, ability, and intersections of social identities. For example, traditional approaches to school mathematics in written and oral English communication place constraints on opportunities for students to experience the beauty of mathematics and demonstrate their own insights. Such limitations of the learning environment can ignore the mathematics capabilities of emergent bilinguals, socially shy students, and students for whom communication is largely nonverbal. Consider inviting alternative forms for expression, representation, and engagement, and encourage students to demonstrate knowledge by creating a podcast, a video, or even a comic strip.

Domain 2: Diversity

Often, simply engaging in the activities that build on the identities of students and families connects the curriculum to the Diversity domain. The goal of this domain is to create a climate that celebrates diversity. To do so, teachers need to recognize that all students are unique and have distinct differences (e.g., differences in student thinking, backgrounds, and cultures) that should be leveraged to strengthen learning for all. Teachers can build on these differences to create a multidimensional classroom that values and leverages many different ways of being mathematically "smart" (Featherstone et al., 2011) as well as students' cultural and linguistic backgrounds. Figure 1.3 describes teacher and student roles that reflect the Diversity domain in the context of mathematics teaching and learning.

---

[1] Through this book and the lessons included, we have intentionally capitalized terms used for people of Color, such as Black and Brown, while leaving white written in lowercase. We follow Frances Harper (2019) in her rationale: "I chose to capitalize Color but not white to challenge the ways that these standard grammar conventions reinforce systems of privilege and oppression" (p. 268).

Figure 1.3. Diversity Domain in the Context of Mathematics Teaching and Learning

DIVERSITY

Teachers:

- Design and implement a curriculum that honors diversity in mathematical reasoning, sensemaking, and multiple forms of engagement to promote individual and collective learning.

- Create multidimensional classrooms, raising students' expectations for contributions from each and every student.

- Deconstruct stereotypes about students' mathematical identities and who can and cannot do mathematics.

Students:

- Express comfort in working with and learning from people who are both similar to and different from them and engage respectfully in collaborative work and discussion.

- Express curiosity about the mathematical contributions and experiences of others and exchange ideas and perspectives in an open-minded way.

- Respond to diversity by building respect, understanding, connections, and empathy for different ways of knowing and being in mathematics classrooms.

Source: Adapted from Learning for Justice (2021).

 Available for download at **resources.corwin/TMSJ-UpperElementary**

As a new teacher, author Cathery Yeh saw an idea in a teacher magazine about creating cultural ABC books. She did this lesson with her ethnically diverse fourth-grade class to start off the fraction unit. An alphabet was sent home with each student, and families assisted in writing a word for each letter in either English or their home language that represented how their family engaged in fraction concepts in their daily lives. Then lessons on unit fractions, equivalent fractions, and fraction operations were developed from the experiences referenced in the ABC book, and families came in to co-teach the mathematics lessons. By developing cross-cultural awareness based on the knowledge and strengths of diverse communities, students can set a foundation of recognizing shared strengths and struggles rather than being derailed by cross-cultural differences and conflict.

Mathematics teaching and learning is complex and embedded in a broader societal context of inequities in which implicit and explicit biases are pervasive. Significant and longstanding disparities in resources, access to rich and relevant tasks, and mathematics learning outcomes continue to perpetuate race, class, language, and ability hierarchies (Artiles et al., 2006; Flores, 2007; NCTM, 2020). Respecting diversity in mathematics requires explicit attention to deconstructing stereotypes about students' mathematical identities and who can and cannot do mathematics. Ability grouping and tracking practices are pervasive in upper elementary school classrooms (Loveless, 2013; National Education Association [NEA], 2015). The research is unequivocal: ability grouping and tracking widen the achievement gap between groups (NEA, 2015) and harmfully affect how students see themselves in relationship to mathematics learning (Boaler, 2015). Teachers who attend to the Diversity domain minimize separated instruction by teaching mathematics in a

By developing cross-cultural awareness based on the knowledge and strengths of diverse communities, students can set a foundation of recognizing shared strengths and struggles rather than being derailed by cross-cultural differences and conflict.

way that accounts for and leverages student differences through an inquiry-based approach. Inquiry encourages students to approach problems in a variety of ways that make sense to them and to bring the skills and ideas they have a sense of ownership of to the task. Infusing options in how students engage with and make sense of the mathematics content allows each learner to gainfully interact with, learn from, and contribute to the learning environment (CAST, 2018). Options can be incorporated into a mathematical activity by giving students choices in how they communicate their mathematical thinking through talking, writing, or drawing and in the availability of visuals, manipulatives, and relevant contexts to build meaning of the mathematical concept. These examples serve as a glimpse into a multidimensional mathematics classroom that supports students in expressing comfort in working with and learning from peers who are both similar to and different from them and valuing difference as strength.

Domain 3: Justice

This domain moves from celebrating diversity to an exploration of how diversity has been used as a marker for oppression that has differently impacted groups of people. Teachers who attend to the Justice domain locate causes of inequalities in social conditions (e.g., tracking, ability grouping, and Eurocentric curriculum) rather than believe that these inequalities stem from innate characteristics of individuals or groups. Figure 1.4 describes teacher and student roles that reflect the Justice domain in the context of mathematics teaching and learning.

Figure 1.4. Justice Domain in the Context of Mathematics Teaching and Learning

JUSTICE
Teachers:
• Locate causes of inequalities in social conditions (e.g., tracking, ability grouping, Eurocentric curriculum) rather than believe conditions are inherent within individuals and can provide students with opportunities to use mathematics to explore these causes.
• Recognize that inequities of the larger society are replicated in common structures and practices that perpetuate disparities in mathematics learning opportunities based on race, class, language, gender, and ability status.
• Explicitly shift the power dynamic between student-and-teacher and student-and-student by centering identities, perspectives, and knowledge traditions that have often been silenced.
Students:
• Recognize stereotypes and pervasive myths around what mathematics is and what it means to know and be good at mathematics.
• Recognize that power and privilege influence relationships on interpersonal, intergroup, and institutional levels and consider how they have been affected by those dynamics in their mathematics learning experiences and in the world.
• Use mathematics as a tool to identify unfairness on the individual, interpersonal, and institutional or systemic level.

Source: Adapted from Learning for Justice (2021).

 Available for download at **resources.corwin/TMSJ-UpperElementary**

Teachers attending to the Justice domain reflect on how inequities in the larger society are replicated in everyday structures and practices that perpetuate disparities in mathematics learning opportunities based on race, class, language, gender, ability status, and other markers of difference. Students are provided with learning opportunities to recognize that power and privilege influence relationships on interpersonal, intergroup, and institutional levels. Students also consider how they have been affected by those dynamics in their learning experiences and in the world by using mathematics as a tool to identify unfairness on the individual, interpersonal, and institutional or systemic level. By helping students to understand how oppression operates both individually and institutionally, they are better positioned not only to understand their own lived experiences but also to develop strategic solutions based on historical and systemic roots rather than romanticized or missionary notions of social change.

The Justice domain is also about liberation. An example is Gloria Gallardo and Cathery Yeh's decimal operations lesson on the Delano Grape Strike, in which students used mathematics to understand the power of community organizing (see Lesson 6.1: *"Tu lucha es mi lucha"*: *Mathematics for Movement Building*). Gloria and Cathery live in California, so they used the local example of the Delano Grape Strike to study the mathematics in movement building. The lesson calls for students to be introduced to the Filipinx labor organizer, Larry Itliong, who led the strike because they understood the importance of their students, mostly Filipinx, knowing the history and contributions of the Filipinx community to U.S. labor movements. However, Gloria and Cathery centered the lesson on the power of intersectional coalitions. When Gloria implemented the lesson with her students, she printed photographs and statistical data from the strikes and protests; students did a gallery review around the classroom, filling out a sheet about their noticings, wonderings, and feelings from their interpretations and analysis of the graphs and photos shared. The activity was followed by a read-aloud from the biography *Journey for Justice: The Life of Larry Itliong* (Mabalon & Romasanta, 2018), allowing students to learn more about Larry Itliong and to see how this important leader served as a member of many movements throughout his lifetime and as part of a broader movement to change unjust conditions. Exposing students to broader movements for justice allows students to identify not only the conditions in which marginalized groups are exploited but also how they came together to fight back.

In an effort to use more inclusive language, we use the term *gallery review* for what is commonly referred to as a "gallery walk."

Domain 4: Action

It's one thing for students to learn about social movements in books and videos and to tell others about it; it's another to participate in creating change firsthand. Figure 1.5 describes teacher and student roles that reflect the Action domain in the context of mathematics teaching and learning.

Figure 1.5. Action Domain in the Context of Mathematics Teaching and Learning

ACTION

Teachers:

- Engage in community- and place-based pedagogies and experiences that bridge mathematics classrooms with community and social movements.

- Understand that learning can emerge from a problem-posing pedagogy, designed around the ideas, hopes, doubts, fears, and questions that occur when students use mathematics to develop "generative themes" about their world.

- Provide students a consistent opportunity to recognize their own responsibility to stand up to exclusion, prejudice, and injustice.

Students:

- Understand the nature and creation of social oppression and feel empowered to intervene and to use mathematics to argue for equitable solutions.

- Make principled decisions about when and how to take a stand against bias and status differences within the mathematics classroom and in their communities.

- Plan and carry out collective action using mathematics as a tool to address injustice in the world.

Source: Adapted from Learning for Justice (2021).

 Available for download at **resources.corwin/TMSJ-UpperElementary**

Drawing from the work of Gutstein (2006) and Freire (2000/1970), students can learn to use mathematics to "read and write the world." Reading the world means using mathematics to develop understanding of a given situation. Writing the world with mathematics involves taking action to create change. In the Action domain, teachers provide opportunities to move beyond raising awareness to supporting students to act on issues that affect them and their communities. Students identify issues they feel passionate about and learn the skills of creating change firsthand using mathematics as a tool to analyze and create change. Students can learn how to improve the material conditions of their lives by learning how to do research, analyze who has the power to change situations, write letters and speeches, use new media (such as blogs, infographics, documentaries, and public service announcements about their cause), and learn other skills with which to work for justice.

The following example incorporates the Justice and Action domains. The fourth-grade teachers at Garfield Elementary engaged their students in a lesson about homelessness and gentrification in their community. They launched the inquiry by inviting students to consider the statement: "On a single night in 2018, roughly 553,000 people were experiencing homelessness in the United States" (U.S. Department of Housing and Urban Development, 2018, p.1). To gain a sense of its magnitude, they read *How Much Is a Million?* (Schwartz, 2004) to envision the number by relating it to quantities mentioned in the book, and then relating a million to their community, such as the number of students in the school, the number of students in the district, and the number of people in the city. Examining the data on homelessness sparked a multifaceted and historical investigation to examine the structural factors (e.g., income, cost of living, affordable housing) resulting in homelessness, particularly for the local community experiencing gentrification. Teachers brought in local community activists from a grassroots organization to show the class how data and statistics are used to highlight gross inequities caused by the city's redevelopment plans. Students and families wrote letters and postcards to local city councils and spoke at board meetings to increase affordable housing units, broaden affordable housing median income guidelines, and increase local residents as meaningful stakeholders in making decisions concerning the development of the land. These actions led to the city's passage of an ordinance designed to increase public participation and transparency in planning and local government (Agarwal-Rangnath et al., 2022).

PAUSE AND REFLECT

- Which of these roles in the four domains do you engage in with the students in your classroom?

- Which of these roles in the four domains would you like to work on in a targeted way to continually improve your practice?

WHY TEACH MATHEMATICS FOR SOCIAL JUSTICE?

When teachers focus only on delivery of mathematics content standards during mathematics instruction, disconnects can occur between the content and students' passions and lived realities. Contextualizing mathematics instruction in students' experiences of social injustice helps them become more interested in mathematics (Rubel et al., 2016). Furthermore, TMSJ supports students' use of mathematics to better understand social injustices they recognize in their lives and to be able to act upon those injustices. In doing so, students learn more mathematics.

So why bring social injustices into the mathematics classroom? Besides offering critical issues for students to examine and serving as context for mathematics development and investigation, five other answers to this question resonate with us. TMSJ helps us to

- Build an informed society,

- Connect mathematics with students' cultural and community histories,

- Connect mathematics to other school subjects,

- Empower students to confront and solve real-world challenges they face, and

- Help students learn to value mathematics as a tool for social change.

Each of these rationales to TMSJ resonates with the foundations of a public education, effective instructional practices, and a purpose of mathematics education shared by mathematics teachers everywhere.

Build an Informed Society

To create a just society, students must become better informed about not only their own lives but also the lives of others that may be different from their own. Students "respectfully express curiosity about the history and lived experiences of others" (Diversity 8) and "respond to diversity by building empathy, respect, understanding, and connection" (Diversity 9). It is paramount that students connect to the injustices expressed by members of their communities—especially to injustices they may be unaware of, as experienced by people with different social and cultural experiences from them. Mathematics serves a special role in informing and educating citizens of these issues. By exploring the context of important issues and relating them to mathematics, students become aware of how mathematics may be used to help them better understand the issue, sorting through misconceptions and rhetoric. A student with a meaningful mathematics education is not just academically successful but also prepared to make informed decisions in a modern, ever-changing society.

Connect Mathematics With Students' Cultural and Community Histories

Too often, students experience mathematics in schools as something detached from meaningful contexts; thus, they perceive it as unfamiliar and unimportant. This leaves many of them with the sense that mathematics is inaccessible and not connected to them or who and what they value. Students bring with them to the mathematics classroom a wealth of informal mathematical knowledge from their everyday cultural and social experiences. In Lesson 5.4 (*Family Story Problems*), for example, students draw on their experiences of eating a meal with family or friends and how food is shared to solve mathematics story problems. These experiences and associated knowledge are valuable resources for mathematics teaching and learning.

We know that when classroom experiences and reasoning are meaningfully connected to students' ways of knowing, the learning that occurs—both cognitively and culturally—is powerful and lasting (National Research Council, 2000). By grounding learning in students' own cultural and community histories, a teacher can create both deeper knowledge and greater valuation of students' own culture.

Connect Mathematics to Other School Subjects

In Lesson 7.1 (*Water is Our Right, Water is Our Responsibility*), students think critically about the Indigenous-led movements for environmental justice in relation to water in our lives, including the impacts of water usage as well as water in relation to the Dakota Access Pipeline efforts. Students can make both mathematical connections to place-value concepts and multiplication but also science connections to ecosystems. This lesson can be interwoven through mathematics and science instructional time. Practices that work to take this approach promote students seeing more authentic and rich connections between the content and their lives. These connections offer opportunities for students to see the ways in which subjects, such as mathematics, are present in other areas of their lives.

Empower Students to Confront and Solve Real-World Challenges They Face

TMSJ helps students understand social oppression and feel empowered to intervene and seek equity, making principled decisions about when and how to take a stand and plan and carry out collective action using mathematics as a tool to understand and address injustice in the world. The teacher's role is to learn about their students to identify themes or issues of interest, and thus help them uncover and explore and confront the issues of injustice their families and communities face. Each lesson in this book includes a section on taking action, providing examples of ways students could work to take a stand and carry out collective action.

Help Students Learn to Value Mathematics as a Tool For Social Change

The potential of education is to support students to create better lives for themselves and a better society for each and every individual. Mathematics is a powerful tool to achieve both goals. When students use mathematics to explore, understand, and respond to social injustices they experience or care about, they learn not only the power of mathematics for social change but also that they are actors on and in the world with the power to transform inequities and create social change. We want students to recognize that their mathematical power can improve the conditions of both their own lives and the lives of others.

REFLECTION AND ACTION

1. Reflect on your teaching. Are there ways you have affirmed students' identities, celebrated diversity, investigated (in)justice, and/or supported students in taking action? How are you hoping to grow in each of these areas?

2. Visit a colleague and initiate a conversation about why they became a teacher. Does the conversation draw on any of these lenses?

   a. Build an informed society.

   b. Connect mathematics with students' cultural and community histories.

   c. Empower students to confront and solve real-world challenges they face.

   d. Help students learn to use mathematics as a tool for social change.

3. Create your own definition for social justice. How might you help students develop their own definition?

CHAPTER 2

FOSTERING A CLASSROOM COMMUNITY FOR SOCIAL JUSTICE

Teaching mathematics for social justice (TMSJ) involves more than teaching a series of lessons at opportune times throughout a school year. In addition to using lessons exploring (in)justice in our world, TMSJ requires the fostering of an inclusive and affirming classroom environment. Teachers facilitate the co-creation of a classroom culture that empowers students and helps them see, recognize, and value how mathematics supports them in understanding, critiquing, and changing their world. Teachers maintain connections with students and their social contexts while also building on students' mathematics and social issue knowledge to co-create new knowledge.

To further consider what is important in the implementation of social justice mathematics lessons (SJMLs), this chapter is organized into five sections: *Context Matters, Content Matters, Who Matters, When Matters,* and *How Matters*. Each section provides guidance for you as you develop a plan to justify the use of SJMLs in your classroom and foster a classroom community for social justice.

- *Context Matters* provides guidance for connecting with your students' experiences and reaching out to families and communities to navigate your current school or district setting as you consider SJMLs that might involve potentially controversial topics in some communities, might result in student responses leading to advocacy or other action, or might raise some parent/caregiver or community stakeholders' concerns or opposition.

- *Content Matters* outlines the importance of making sure that SJMLs focus on essential mathematics content, are part of a coherent learning experience for students, and balance the focus on both mathematical and social justice goals.

- *Who Matters* provides guidance for navigating assumptions that TMSJ is only or primarily about helping traditionally marginalized students see themselves in the curriculum. TMSJ is about learning mathematics through understanding the systemic oppression that affects us all and thus everyone, from positions of privilege and oppression, working to dismantle oppressive structures.

- *When Matters* considers the experiences and readiness of elementary students to engage in these topics and provides recommendations for ways to integrate SJMLs with other content area learning goals and classroom routines.

- *How Matters* highlights the fact that the lessons provided in this book have an inquiry-based, or exploratory, approach and supports you in choosing pedagogical strategies that will engage each and every student in the lesson. Further, this section provides guidance for anticipating and responding to students' myriad of reactions to each SJML.

We present these five ideas and encourage you to consider them as you begin to plan the implementation of SJMLs in your upper elementary classroom.

CONTEXT MATTERS

As a preview to this section, Figure 2.1 lists some guiding questions for you to ask yourself or your team to begin to consider how context matters when designing and implementing lessons that address social justice issues.

Figure 2.1. Guiding Questions for Context Matters

Context Matters
When considering any SJML, ask yourself: • What issues are of importance to students, families, and community members? • What do I know about this social inequity/injustice and what do I need to know that intersects with local concerns and interests? How might this topic be received in my local setting? • What is my purpose for including the social justice topic as part of instruction? • Who is on my team of allies ready to support me? How will I communicate both the social justice and mathematics goals to parents/caregivers, administrators, and other stakeholders? • How does the SJML contribute to building students' mathematical and/or social agency and identity? How might this SJML allow students to share and develop an understanding of inequity or a sense of empowerment and liberation?

 Available for download at **resources.corwin/TMSJ-UpperElementary**

A key purpose of schooling is to help young people become informed and able contributors to our society as well as critically thinking members of our democracy. As such, topics of social and political relevance must be a part of the curriculum. Topics that seek to help students gain an understanding of inequities in their school, community, or society, and which teach students to advocate for change, may draw attention from those who wish to avoid controversy or seek to maintain the status quo. Topics themselves are not inherently controversial; what is considered controversial varies based on time, place, people, and the local, national, and global sociopolitical contexts. So, when we name a topic as controversial, we really mean to indicate that it has the potential to create controversy or disrupt a space of comfort (for some) or the status quo.

As you consider the context of your setting and classroom, it is important to work to know and understand the climate and culture of your school and communities in regard to the teaching of controversial topics, especially those that people might not see as traditionally fitting into elementary mathematics curriculum. Who students are, what they know, what they and their families have experienced, and what they care about matter in the classroom (Steele & Cohn-Vargas, 2013). Elementary-aged students already possess experiential knowledge on issues related to identity, fairness, culture, and justice. Parents, grandparents, aunts, uncles, friends, cousins, caregivers, neighbors, and community members share stories about their lived experiences and perspectives—both about mathematics and social issues—and students carry this knowledge inside of themselves. Thus, it is particularly powerful to work *together* with students, families, and communities to identify topics to help students gain new or further understanding of injustice in their school, community, or society.

In addition to improving relevance and honoring local knowledge, one overall goal in carefully connecting to context is to help plan and anticipate possible opposition and resistance that might come up during or after the lesson (e.g., parent/caregiver complaints to administrators, students forming some type of protest, students having their feelings hurt). Thus, as you plan for SJMLs, you need to consider recommendations, such as those in Figure 2.2, when pursuing topics of interest to students, particularly those that may be controversial in your community.

Figure 2.2. Considerations for Pursuing Controversial Topics in Upper Elementary Classrooms

Considerations for Pursuing Controversial Topics in Upper Elementary Classrooms

1. Consider what you have done (thus far) to create an open, inclusive, respectful learning environment with your students.

2. Consider what you have done to incorporate family and community knowledge into the curriculum (e.g., using community reviews* or surveys, guest speakers, video projects, dialogue).

3. Check to see if your school or district has policies or procedures for teaching controversial topics (e.g., rights of students who identify as transgender, politics, religion).

4. Look at cross-curricular connections to connect social justice goals with learning goals in literacy, science, social studies, art, health, and so on.

5. Identify community stakeholders (e.g., parents/caregivers, businesses, religious groups) who work with related issues and consider inviting them to be resources during lesson planning or teaching.

6. Be mindful of social injustices that are recent hot-button issues in your school, district, or community.

7. Determine if the social injustice is one that requires discussion with your administrator before teaching.

* In an effort to use more inclusive language, we use the term *community review* for what is commonly referred to as a "community walk."

 Available for download at **resources.corwin/TMSJ-UpperElementary**

To further think through your context and setting, also consider your reason for introducing the social injustice during mathematics time. You need to consider your personal biases, creating a climate of success, your allies, and the timing for the SJML you have selected.

Bias

We all have biases whether we are conscious of them or not; we are influenced by our feelings, opinions, backgrounds, and prior experiences. We have biases because we are human, and power, privilege, and oppression operate both individually and institutionally. Reflecting on our bias allows us to be better positioned to not only understand our own lived experiences but also to develop strategic solutions based on historical roots rather than romanticized and missionary notions of social change. Therefore, preparing for a SJML requires consideration of personal and institutional biases and background knowledge regarding the topic of the lesson and how it may affect students. For example, when leading Lesson 5.1 (*Families Matter*) on family structures, it is critical for all teachers to consider their own family structure and their experiences and exposure to the diverse family structures that make up the classroom and school community. Teachers should also consider common narratives of the nuclear family (e.g., mom, dad, brother, sister, and dog) that are pervasive in social media and children's books, influencing our own and students' perceptions of what a family should and can be (e.g., multiracial, adoptive, LGBTQIA+, multigenerational, and single-parent families). There is invisibility and avoidance when we have identities of privilege (e.g., white, Christian, male, heteronormative, able-bodied) to understand the challenges, resilience, and strengths of traditionally marginalized communities in the United States.

Unpacking biases often requires learning about oneself and the experiences of others. This includes identifying and unpacking the implicit ways you navigate the world and extends to a consideration of how your students' experiences may differ from yours and how you will navigate that space. A white teacher exploring racism with a racially diverse group of students, for instance, should also take special care to consider what and how they will communicate to their students both explicitly and implicitly about where they stand on a topic and to center on the experiences of the group. Invite family and community members to share their life stories. Go out into the community to get to know diverse neighborhoods and cultural traditions. Attempting to take a "neutral" stance as opposed to one of anti-racism in the face of injustice may be harmful to your students and lead to distrust and an unwillingness to engage with the content.

Respect

Create a climate of respect for diversity. Encourage students to share their knowledge and experiences about their own cultural backgrounds and social identities with their classmates. A sense of pride in their culture, heritage, race and ethnicity, skin tone, and gender is cultivated when teachers provide students opportunities to learn about and share out who they are and where they are from. When students are supported to learn more about their own identities and cultural history, they are better able to identify, deconstruct, and challenge harmful stereotypes about their identities and to operate from a place of pride about their communities. Having students learn to listen with kindness and empathy to the experiences of their peers allows the class to deconstruct stereotypes and learn more about the history, strengths, and resilience in each other's cultures. By developing cross-cultural empathy based on historical knowledge and strengths of diverse communities (see Lesson 6.1, *"Tu Lucha es mi Lucha": Mathematics for Movement Building*), students can set a foundation of recognizing shared connections for movement building rather than being derailed by cross-cultural conflict.

Audience

Think about the students in your classroom and those with whom they will interact in other classes, extracurricular activities, and outside of school. Controversial topics have a way of reaching others, sometimes like rumors or bad press and other times as respectful curiosity about the world in which we live. With your students, co-construct classroom norms so that all are comfortable expressing their individual perspectives on issues without fear of personal attack from their peers (see

Figure 2.3). These norms should support all students in "respectfully expressing curiosity about the lived [social and mathematical] experiences of others" and to "exchange [mathematical and other] ideas and beliefs in an open-minded way" (Diversity 8; Learning for Justice, 2016). It is also important to recognize that levels of comfort are not the same for all students. For some students, their social identities and perspectives have been in the margins of schooling, and thus care must be taken by teachers to not only hear from all students but to also validate historically marginalized social identities and perspectives. It is also important to ensure that students understand that they and their peers are speaking for themselves and are never expected to represent an entire group (such as all Black people). The art of teaching issues related to inequities involves co-establishing an identity-safe classroom where students believe their social identities are assets and diversity is a resource (Steel & Cohn-Vargas, 2013) as well as where acknowledgment and respect are a part of the culture. It is important to remind students of the value of seeking to understand so they can "express comfort with people who are both similar to and different from them and respectfully engage with all people" (Diversity 6) and "respond to diversity by building empathy, respect, understanding, and connection" (Diversity 9) (Learning for Justice, 2016).

Figure 2.3. Possible Classroom Norms for SJMLs

Possible Classroom Norms for SJMLs

1. We deeply listen to what others say and to feelings, experiences, and wisdom behind what they say.
2. We recognize that other people may hold pieces of the puzzle that we don't.
3. We use personal experiences to share our ideas about the topic.
4. We do not necessarily always agree; thus, our goal is to gain deeper understanding.
5. We can provide support (e.g., head nods, smiles) when someone is struggling to speak.
6. We speak for ourselves as individuals and do not make assumptions about others' experiences.
7. We disagree with ideas, not people.

 Available for download at **resources.corwin/TMSJ-UpperElementary**

Allies

Students, parents/caregivers, community members, colleagues, and administrators are all people whom you should consider when seeking allies in the planning process. Remember that students have knowledge and experiences and many of them know a great deal about the social context you are considering for the lesson. They may even know more than you about the school and community context, and they may know individuals who might be allies in planning. Also, consider inviting family and community members who may work in a field related to, or have experience with, the social injustice to come in and observe your classroom and provide feedback. Similarly, consider providing your administrators and/or colleagues a brief overview of your lesson plan and collaborating with them on the planning and implementation of the lesson(s). You will be less open for criticism when you have collaborated with others.

PAUSE AND REFLECT

Ask yourself the following questions:

- Who is on my team of allies ready to support me?

- Have I been open to listening to the voices of others, or am I steamrolling ahead with what I think is a good idea?

- How can I learn more about my students, their families, community members and organizations, colleagues, and administrators to build relationships and create allyships?

Timing

As the saying goes, "timing is everything," so consider when the SJML will be taught. On the one hand, choosing to discuss a topic immediately after a related incident in your school, district, or community (e.g., police brutality) may not be the most opportune time for the lesson. For example, it may not be best to engage in a lesson on racialized violence shortly after a student of Color has been attacked in your school or community. Waiting and allowing students time to process what has happened might have more merit. On the other hand, students remember the silence (Dunn, 2021), and not discussing a current event may communicate a lack of importance and/or avoidance of a topic students are already thinking about and affected by. Using a current event may very well be a powerful way for students to process the issues and for you to center the students and humanize them in the classroom. In some cases, a middle ground of discussing an important issue immediately after it occurs coupled with a more in-depth mathematical lesson later may be appropriate. This decision relies strongly on consultation with your allies and your students as well as on your personal expertise.

CONTENT MATTERS

As a preview to this section, Figure 2.4 lists some questions for you to ask yourself or your team to begin to consider how content matters when designing and implementing lessons that address social justice issues.

Figure 2.4. Guiding Questions for Content Matters

Content Matters

When considering any SJML, ask yourself:

- How will the lesson contribute to the learning goals for my class?
- How does this lesson contribute to developing students' deep understanding of mathematics? How does the lesson empower students mathematically?
- How does the lesson connect to an issue that is relevant to my students?
- How does the lesson promote anti-bias education by addressing prejudice reduction and/or collective action? Which domain of the Social Justice Standards are addressed? Which Social Justice Standard?
- How does the SJML allow students to use mathematics as a sociopolitical tool of analysis?

 Available for download at **resources.corwin/TMSJ-UpperElementary**

We recognize that each lesson's content plays a major role in helping students make sense of both mathematics and their world by using mathematics in an authentic and empowering way. When selecting (or designing) a SJML, you should make sure that the lesson's content contributes to helping you achieve:

1. The overarching goal of developing students who think critically as doers of mathematics and exchange curiosity about the mathematical contributions of others;

2. Mathematics goals related to required or recommended grade-level content standards; and

3. Deeper understanding of issues relevant to your students' lived experiences *and* how students might advocate for change.

When you think about how the content of each lesson contributes to the mathematics story for your students, the careful selection (or design) of lessons and associated tasks is critical. Lessons should be inquiry based and should support students in making mathematical connections and deepening their understanding of mathematics.

Planning a SJML requires you to not only be thoughtful about the design and use of mathematics but also to think deeply about the context of social injustice. There is often a tension when designing or implementing SJMLs between the mathematical goal and the social justice goal, where one may be foregrounded to the detriment of the other. You may feel nervous that you are not an expert in certain topics—or may not feel expert enough. Part of your preparation should be learning both about the mathematics content and the social (in)justice topic. For the mathematics, engage in the mathematics yourself, consider how students might approach problems, and ask yourself how the task or mathematical activity supports diversity in mathematical reasoning. Here are some questions for you to consider: What mathematics will students need to understand and examine this

> Planning a SJML requires you to not only be thoughtful about the design and use of mathematics but also to think deeply about the context of social injustice.

issue? What data or mathematics will they need to support conclusions, and why? For the social (in)justice topic, learn all you can, not just from media and alternative media outlets but also from people closely involved who can provide the most insight. With respect to the integration of mathematics and social justice in your SJML, remember that for both mathematics and the social (in)justice topic, one lesson is not sufficient for full understanding. Consider your SJML goals in relation to how you hope to adequately contextualize the social issue and develop mathematical understanding over time (Bartell, 2013). In any case, you will certainly not know everything there is to know or you may stumble on sensitive issues or an unexpected mathematical connection that a student makes. Engaging in social justice work allows us to move away from being experts to learning with our students and the community. It is only when we're together that we can notice things that we did not when alone. You can position students as competent mathematical and human beings and collaborators (or co-conspirators; Love, 2019) in the content and the social (in)justice topic and ask to learn together.

Having examined both the mathematics content and the social (in)justice topic, implementing a SJML implies that at the forefront of planning, you

- View students as competent mathematical beings in which their lived experiences and community and cultural ways of knowing are leveraged during mathematics instruction;

- Design lessons that help students see that they are knowers and doers of mathematics and which deconstruct negative stereotypes about who can and cannot do mathematics;

- Provide opportunities for students to use the mathematics they know and further grow their mathematical skills;

- Seek out opportunities to infuse topics that are relevant to your students' background, culture, and lived experiences; and

- Guide students as they use mathematics as a tool to identify unfairness on the individual, interpersonal, and institutional or systemic levels.

When preparing the content of a SJML, you must also consider the age appropriateness of both the mathematics and the social justice content. Your state standards document can likely serve as a guide to the appropriateness of the mathematics content. With respect to the social injustice, research documents that young children aged 3–5 years have racial awareness (Goodman, 1952) and connections to ideas of fairness in elementary grades can support anti-bias curriculum in the classroom (Mistry et al., 2017). Children aged 9–12 years begin to understand historical and geographical aspects of identity and a deepening awareness of cultural/political values (Derman-Sparks et al., 2020). Students rely on teachers (or other trusted adults) to help them make sense of the confusing messages the world sends them about race, fairness, diversity, and justice. For example, the racial stereotypes of how Asian and white people are supposed to be good at mathematics, and by association then Black, Latinx, and Native American people are not, starts early, with some research suggesting that students are aware of such academic stereotypes by second grade (Cvencek et al., 2011). Our recommendation is that you speak with trusted friends, students' families, community member allies, and possibly experts on the age appropriateness of the social content.

At any age, some information or discussions may be traumatizing due to students' personal experiences with the injustice. Be prepared to respond to this in humanizing and empowering ways; an important first step is to validate feelings and experiences. Be sure to follow up with a school counselor or specialist to ensure the student has an opportunity to find support.

WHO MATTERS

Figure 2.5 lists some questions for you to ask yourself or your team to begin to consider who you are designing and implementing SJMLs for and with.

Figure 2.5. Guiding Questions for Who Matters

Who Matters
When considering any SJML, ask yourself: • How are the students in my classroom similar to and different from one another across identity groups (e.g., race, class, gender, religion)? • How does this lesson engage students in respectfully exploring the historical, lived, and mathematical experiences of themselves and others so as to exchange beliefs, ideas, and perspectives in an open-minded way? • How does this lesson support each and every student in expressing pride, confidence, healthy self-esteem, and positive mathematical identity without denying the value and dignity of other people?

 Available for download at **resources.corwin/TMSJ-UpperElementary**

Who your students are, in a myriad of ways, should inform your SJMLs. Many people believe TMSJ is only or primarily about helping traditionally marginalized students see themselves in the curriculum, but this is not the case. TMSJ is about *all* students learning mathematics through understanding and working to dismantle systemic oppression and injustice. It is important, for example, for affluent students to learn about injustice and social justice reform as many, although they express concern about inequities, initially connect the problem to individual shortcomings rather than systemic disadvantages (Swalwell, 2013). Whether intentional or not, you likely hold stereotypes or biases about some of your students and the knowledge that they bring to the classroom. Students, too, likely hold stereotypes or biases about other students.

TMSJ is about *all* students learning mathematics through understanding and working to dismantle systemic oppression and injustice.

The Social Justice Standards in the Justice domain can inform your selection and implementation of SJMLs. Think of who your students are, and in your classroom engage in conversations that support students to

- Get to know people as individuals because they know it is unfair to think all people in a shared identity group are the same (Justice 11)

- Know when people are treated unfairly and give examples of prejudiced words, pictures, and rules (Justice 12)

- Know that words, behaviors, rules, and laws that treat people unfairly based on their group identities cause real harm (Justice 13)

- Know that life is easier for some people and harder for others based on who they are and where they were born (Justice 14)

- Know about the actions of people and groups who have worked throughout history to bring more justice and fairness to the world (Justice 15)

It is critical that you understand who your students are by working to understand each and every student as a whole child, including the world that the students are a part of outside of school. Knowing students well and deeply can keep you from succumbing to the reputations and biases some students carry that are assigned from other students, families, and colleagues.

Understanding each and every student as a whole child includes recognizing that some students have experienced trauma, and at times, school has perpetuated the trauma. Trauma can be defined as the reaction to a shocking or painful event or series of negative events. With TMSJ, it is important to engage without retraumatizing students. You can take some specific steps to mitigate the effects of trauma for students in your classroom.

Know Your Own Story

Teacher educator Yolanda Sealey-Ruiz describes how teachers should "dig deep and peel back layers of themselves and think about how issues of race, class, religion, and sexual orientation live within" (Dillard, 2019). In particular, in the context of the United States, our racial wounds run deep—through all of us. It's important as teachers that we have spent time working on understanding and healing ourselves from our racial trauma if we expect students to be open and honest with us in our classrooms (e.g., see Menakem, 2017). In addition, identify and confront your own biases—about yourself, your students, colleagues, communities, and the world. What is beneath these layers that you peel back will affect relationships with your students. Worse, if these issues go unexamined, they may cause harm.

Learn From and for Your Students, Their Families, and Their Communities

It is important to learn both from and for your students, their families, and their communities (Pitts, 2016). Learn *from* your students, their families, and their communities: How are your students brilliant? What are your students' needs? What are their families' needs? What are their community's needs? What are their struggles? Also, learn *for* your students, their families, and their communities: Read literature to further your understanding of issues your students, their families, and their communities face. Connect yourself with others who are doing this work to learn together. Increase your knowledge about trauma and how it may manifest for your students. For example, learn about how the curriculum you may use in your own classroom can be violent (even if you do not intend it to be), causing students a type of emotional destruction legitimized as teaching. You can then prepare to avoid curriculum violence, as you have learned from your students, their families, and their communities, by adopting an anti-racist framework and pedagogy, for example (see *Ending Curriculum Violence* by Jones, 2020).

Establish Social and Emotional Support

Students who have and are experiencing trauma in their lives (or in their schools, including their mathematics classrooms) have their sense of safety compromised. You can support students by working to establish a socially and emotionally supportive classroom environment where you teach and model empathy, intervene in hurtful exchanges or prejudice and injustice, and establish connectedness with anti-biased and community-building curriculum and classroom routines.

When preparing the content of a SJML, explicitly consider what you do and do not know about your students and how the SJML may, unintentionally, retraumatize some of them. Do not let fear of harm stop you; rather, work together with allies and engage in the steps described earlier to mitigate the effects of trauma for students in your classroom.

WHEN MATTERS

Figure 2.6 lists some questions for you to ask yourself or your team to begin to consider when to implement SJMLs.

Figure 2.6. Guiding Questions for When Matters

When Matters
When considering any SJML, ask yourself:
• How does this SJML build on previous lessons or preview future lessons, both with respect to mathematics and social justice goals?
• How does this SJML integrate learning goals across content areas (e.g., social studies, science, literacy)?
• What might my students already know about the mathematics and social justice topic of this SJML?

 Available for download at **resources.corwin/TMSJ-UpperElementary**

Social justice mathematics cannot wait until students are "older." As we have argued, upper elementary students are both capable of and interested in engaging with issues of social justice and engaging in advocacy. One of the goals of this book series is to lay a pathway from early childhood through high school to help students develop their mathematical agency in the world, to be critical creators of the mathematics needed to understand the world, and to be empowered to take action. Our hope is that as students grow up in a world where from a young age they are encouraged to think interdisciplinarily, to connect issues of fairness to empathy and understanding, and to advocate for themselves and others, they will be strong advocates for justice and well versed in the role mathematics plays in social action.

> **Social justice mathematics cannot wait until students are "older." As we have argued, upper elementary students are both capable of and interested in engaging with issues of social justice and engaging in advocacy.**

Upper elementary students bring understandings of mathematics to the classroom that can aid in making sense of big issues, while they also learn more mathematics. In our experiences, we hear from teachers that upper elementary students may be interested in the topics, but they do not have the appropriate mathematics skills in their toolbelt to engage in multistep mathematics problems on real-world topics. But the truth of the matter is that they do, when the lesson is planned around mathematical sense-making that leverages their prior knowledge and their vast wealth of community and cultural knowledge (e.g., their lived experiences).

It is important to take up issues of social justice within mathematics instructional time. This helps to strengthen the experiences students have with mathematics and social justice as an integrated area, as opposed to separating it out in some attempt to keep mathematics neutral. It is very possible that not every mathematics lesson in your classroom will have a strong social justice component; however, across a school year, many lessons can and should.

To aid teachers in this process, we advocate for blurring the boundaries between types of content lessons. Many of the lessons in this book incorporate literacy, science, and core mathematics content in the service of exploring and acting on a social justice issue. In addition to being good lessons to try, teachers can reflect on how lessons are structured to be interdisciplinary as a way to improve one's own lesson design skills.

HOW MATTERS

As a preview to this section, Figure 2.7 lists some questions for you to ask yourself or your team to begin to consider how you might implement lessons that address social justice issues.

Figure 2.7. Guiding Questions for How Matters

How Matters
When considering any SJML, ask yourself: • What pedagogical strategies will I use to engage each and every student in the lesson with both the mathematics and social justice topic? • How am I reflecting on and addressing my own bias and positionality to create an anti-bias mathematics classroom? How am I supporting students in reflecting on and addressing their bias and positionality? • What questions will I use to facilitate students' learning of both mathematics and social injustice? • How might my students react to this SJML, and how will I prepare for that?

 Available for download at **resources.corwin/TMSJ-UpperElementary**

The SJMLs in this book have an inquiry-based, or exploratory, approach and employ several different instructional models. We encourage you to thoroughly review each lesson, its instructional model, and the lesson goals (mathematics and social justice) to determine the appropriateness for use in your classroom, adjust as needed in order to localize the lesson so that it is relevant to your students, and plan for the appropriate level of guidance and support. Next, we present a SJML Planner (Figure 2.8; see also Appendix F) that the series team developed, attending to *Context, Content, Who, When,* and *How,* to help you organize information in your initial preparation for a lesson. Please note that the SJMLs in this book do not provide information on each of these sections, nor are they structured in this form. Some of the decisions that you need to make when planning are specific to your local and personal contexts. While this template can help with the planning that is needed when introducing issues of social injustice during mathematics time, additional questions are also provided in Figures 2.1, 2.4, 2.5, 2.6, and 2.7 to help you think through the elements of the SJML Planner. We close this chapter by providing guidance for anticipating and responding to myriad reactions of students and parents/caregivers to each SJML.

RESPONDING TO WONDERINGS

As SJMLs are launched in upper elementary classrooms, students may generate more wonderings as they begin to grapple with their own interpretation of social issues as well as ways in which the narratives they may have heard outside of schools are different from those they are examining. Students may openly share their own wonderings and reactions to the issues being examined with mathematics, along with those expressed by parents/caregivers in their lives. As wonderings are brought up in the classroom setting (or beyond), think positively—the student trusted you enough to bring them up. They see these conversations as valuable; your classroom is a safe place to ask questions and seek answers. Follow up on these wonderings of students by asking for more information and finding resources for them to explore answers, including partnering with communities and families.

Figure 2.8. Social Justice Mathematics Lesson Planner

Social Justice Mathematics Lesson Planner

PART I

CONTEXT
Purpose
Audience
Allies
Timing

CONTENT	
Mathematics Goal(s)	**Mathematical Content Domain**
Social Justice Topic and Brief Description	
Social Justice Outcome	

WHO	
Resources for Your Learning	Classroom Practices and Norms to Establish Social and Emotional Support
How Your Lesson Supports Students in Recognizing Injustice at Both Individual and Institutional or Systemic Levels	

WHEN
Possible Interdisciplinary Connections

HOW
Instructional Model (e.g., Classroom Routine, Mathematics Task, Three-Act Task)

PART II

Launch/Engagement	
What will the teacher do?	**What will students do?**

Exploration/Investigation	
What will the teacher do?	**What will students do?**

Summarize/Discuss	
What will the teacher do?	**What will students do?**

Taking Action

Stakeholder Communication

How is the teacher communicating lesson goals?	How are students communicating their learning?

RESPONDING TO PUSHBACK

Planning for a SJML requires you to think about more than the mathematics goals; you also need to consider the social justice goals and thus the social injustices that might be addressed in the lesson. This chapter was designed to support you in your planning process. The following paragraphs outline some recommendations for responding to pushback.

- *Seek to understand others' perspectives.* Ask questions to gain an understanding of any raised issues or concerns (e.g., Why do you say that? What would you like me to know more about?). Make a mental note of any power dynamics, privileges, or biases that might be at play. For example, how might status dynamics between students (with respect to mathematics, social status, and others) position one student as "knowing" and another as "off task?" (For a brief discussion of implicit biases in mathematics classrooms, see Barshay, 2018.)

- *Notice when you are responding from an emotional stance.* Monitor the emotional levels and try to take a break in the conversation when dealing with heightened emotions. Notice when you are responding from an emotional stance, name it, and decide if that means you need to take a break from the conversation versus work through the issues in the moment.

- *Share the rationale for integrating social justice topics into mathematics instructional time.* Use the guiding questions (Figures 2.1, 2.4, 2.5, 2.6, and 2.7) and the Social Justice Mathematics Lesson Planner (Figure 2.8) to prepare for the lesson and anticipate outcomes or reactions. Having a written plan to reference can show the thinking that went into the lesson planning.

- *Avoid sending lengthy responses through email.* People have very little control over the perceived tone of an email and how parts of a message will be interpreted or used. A timely response is needed, and keep it brief. For example, "Thank you for your email. When is a good time for me to call to discuss your concerns?"

- *Pick battles carefully.* Recognizing areas of compromise and noncompromise in the conversation is important. Refrain from getting caught up in small nuances and details. Remember that the goal is to continue to use SJMLs during mathematics time, so retreat to teach another day when the opportunity presents itself.

CONCLUSION

For many of us, the implementation of SJMLs in the upper elementary context is new. Much of what is required, however, is already in your repertoire. You have already implemented many aspects of TMSJ, such as organizing your curriculum around content standards and teaching in inquiry-based ways. You use many strategies to engage students in meaningful discourse centered on mathematical ideas.

You draw upon students' social, cultural, and academic resources in your instruction. For many of you, the aim to develop a critical consciousness in your students is something you value and are ready to dive into. The SJMLs in this book can be an excellent starting point.

In this chapter, we asked you to consider five elements in preparation: *Context, Content, Who, When,* and *How.* Here are some questions to consider: What social justice issues are of importance to students, families, and community members? Who is on your team of allies ready to support you? How will the lesson contribute to developing students' deep understanding of mathematics and/or empower students mathematically? What instructional strategies will you use to engage each and every student in the lesson, both the mathematics and social justice topics? Many of us feel expert in teaching, and sometimes expert in mathematics, but are quite wary of trying to teach about social injustice. Our experiences suggest that a good approach is to think ahead and learn about the mathematics and social justice issue, be open to learning more from and for students, and foster a classroom environment that both models empathy and views students as competent mathematical beings.

REFLECTION AND ACTION

Consider the following strategy for collaborating on the implementation of one of the lessons in this book, or for developing your own. Working with a small team, organize your discussion using the following protocol:

1. Individually, write down a topic of injustice that may be of interest to your students on a note card or slip of paper. Create an additional note card for each topic. Spend several minutes identifying topics of interest for follow-up.

2. Place all the cards in the bag, then draw out one card. Discuss this topic as a team.

3. Begin discussion by responding to the *Context* and *Content* questions from Figures 2.1 and 2.4.

4. Use all components of Part 1 from Figure 2.8, the Social Justice Mathematics Lesson Planner, to begin planning a lesson that might be used for this topic. (This begins to address the *When* and *How* questions in Figures 2.6 and 2.7).

5. For your topic, identify any allies needed before teaching the topic. What information would each of these people want to know? What might be learned from them?

6. Pair up and role-play with a peer on how the topic/lesson idea can be shared with an administrator or colleague.

7. Finally, debrief with your team. Which section (*Context, Content, Who, When,* or *How*) was most difficult to complete?

CHAPTER 3

INSTRUCTIONAL TOOLS FOR A SOCIAL JUSTICE MATHEMATICS LESSON

We know that you already have pedagogical approaches that promote robust mathematics teaching and learning, or that you know where to go to learn new approaches. As you think about establishing a classroom environment and culture based on teaching mathematics for social justice (TMSJ) and integrating issues of social justice (and the four domains of Identity, Diversity, Justice, and Action) into your classrooms, we recommend that you take some time to reflect on your pedagogical strengths, consider areas you would like to enhance, and identify new pedagogical approaches you might want to learn. The goal of this chapter is to build on and enhance existing resources, leveraging the tools you are already using to embark on your journey—new or continuing—into implementing social justice mathematics lessons (SJMLs). In this chapter, we provide you with a brief review of pedagogical approaches in upper elementary settings, attending to the tenets of each that can provide a pathway to SJMLs. Part of this pathway includes exploration of critical areas included in the high school edition of this book—goals, assessment, and instruction—while attending to the uniqueness of upper elementary settings in humanizing ways that honor students' voice, agency, and identities.

PEDAGOGICAL APPROACHES IN UPPER ELEMENTARY SETTINGS

Enter some upper elementary classrooms and there will be a buzz about the room. Students will be engaged in conversation with each other, working with various mediums to create representations and to make sense of their world. Educators in upper elementary settings often describe their pedagogical approach to teaching as inquiry-based instruction or problem- or project-based learning.

Inquiry-Based Instruction

Teachers using an inquiry-based approach in their classroom acknowledge students' natural curiosity and lived experiences. Based on this, they design learning experiences that provide an opportunity for students to develop new knowledge through exploration and discovery. Inquiry-based mathematics instruction actively engages students in solving mathematics tasks by working collaboratively and using multiple representations to solve problems. Students have opportunities to share and clarify their ideas with one another as well as to make their thinking visible for consideration and critique by other students. Instead of a teacher presenting facts or procedures, the teacher in an inquiry-based classroom encourages students to talk about a problem and draw on their funds of knowledge to understand it. By navigating the direction of their learning and how they will seek out information, students begin to determine the more efficient ways in which they can learn information, taking an active role to collect and retain new knowledge (Nell et al., 2013). The processes of questioning and collecting and making sense of information involve active, hands-on experiences aligning with instructional approaches that value interaction, discovery, and active engagement in learning.

> **The processes of questioning and collecting and making sense of information involve active, hands-on experiences.**

Problem-Based Learning

Problem-based learning begins by posing a problem to students and supporting them in their exploration, generation, and presenting of possible solutions. Problems drive the learning process. As students work to solve mathematics tasks, they gain new mathematical knowledge through opportunities to think critically, present ideas, and communicate with others mathematically. The same can be said about acquiring deeper knowledge and education of real-world issues. SJMLs in a problem-based learning context engage students in using active exploration of real-world challenges with social issues. Problems typically center on a single subject with solutions communicated via presentation or another written element.

Project-Based Learning

Project-based learning rests on the three pillars of rigor, relevance, and relationships (Galindo & Lee, 2018). Project-based lessons begin with an entry event to captivate students and support their recognition of the connection between the work they are about to embark on in the real world and mathematics. Projects typically draw from multiple subject areas and often involve community elements in the form of expert speakers or professionals who provide insight into the topic being explored. At the culmination of the project, students are typically responsible for sharing a product or artifact (project) they have developed in response to the topic posed.

PAUSE AND REFLECT

Consider the following:

- What pedagogical approaches am I currently using in my own classroom?
- How might I work to center mathematics and social justice in these approaches?
- How will I communicate this to colleagues, administrators, and families?

PREPARING SJMLS

Preparing SJMLs may seem new, daunting, exciting, or a range of things at first. We find that reminding ourselves that working as collaborators, or co-conspirators (Love, 2019), with students, their families, and their communities helps calm any nerves. As co-conspirators, Love calls on us, as educators, to shift from supporting and standing by the side of those we believe in, to doing the work ourselves, reflecting and actively leveraging our privilege to dismantle institutional oppression. We, as educators, do not need to know it all; we just need the quality of humility to listen and dialogue with students, families, and communities. Developing SJMLs with students and families allows us to bridge mathematics classrooms to community and social movements and silence critical voices (in or outside of our heads) that aim to perpetuate the narrative that teaching is neutral. Hopefully, from reading the prior chapters, you can see that teaching is not, in fact, neutral; rather, teaching is layered with decisions deeply rooted in our identities and politicized decisions often made in, outside of, and about schools. In this section, we provide insight into the preparation of lessons and the planning process. These revolve around ideas such as establishing goals, identifying and addressing mathematical content and practices, and attending to social justice standards.

Establishing Goals

What do I want all students to know and be able to do?

It is important to consider the first component of the work of teaching a lesson: planning. A critical starting point for planning is to identify clear, profound learning goals. The goals established for your students guide the decision making about task selection and implementation; how you will assess your students; and, ultimately, how your students will see themselves and their peers as confident critical thinkers and doers of mathematics, using mathematics to understand and critique the world. The lessons in this book are unique to upper elementary mathematics classrooms because they explicitly identify social justice outcomes and standards for mathematics content and practices for this grade band. Next we discuss the considerations necessary to attend to both social justice and mathematical content learning outcomes.

In the SJMLs in this book, teachers' goals were strategically designed for their unique groups of learners based on both the prior experiences of the students and possibly students' own input based on interest or area of need. You will see that when designing SJMLs, establishing mathematics goals is important, but of equal importance is the development of social justice goals that range from the development of awareness to taking an active role in addressing instances of injustice.

In Lesson 5.1 (*Families Matter*), lesson author Nicky Meindl takes a critical look at how families are often portrayed in children's literature, a springboard commonly used in many mathematics classrooms to foster mathematical exploration. Within this lesson, students work on the mathematics goals of drawing graphs to represent data and solve problems. At the same time, teachers are facilitating conversations around social justice goals centered on family structure and diversity versus normative representation. Part of this conversation addresses the assumption that all family structures are similar, as teachers pose questions and introduce examples of a variety of family structures and ask students to analyze the literature in their classroom for how family structures are represented.

Mathematics Content

What important mathematical ideas do I want students to understand and be able to use?

"When mathematics instruction goes deep, children are empowered to explore the richness of the mathematical landscape" (Huinker, 2020, p. 77). Elementary instruction should build a strong foundation of deep mathematical understanding, emphasize reasoning and sensemaking, and ensure the highest-quality mathematics education for each and every child. The lessons in this book include the five content domains highlighted in the NCTM book *Catalyzing Change in Early Childhood and Elementary Mathematics* (Huinker, 2020). In *Catalyzing Change*, NCTM identifies research-informed pedagogical approaches to teaching mathematics content deeply honoring important findings from research on students' mathematics learning and recommendations from professional organizations. Figure 3.1 provides a list of the five mathematical content domains and the types of opportunities needed to build a deep foundation of mathematical understanding for each content domain (Huinker, 2020; see Appendix C).

In Lesson 6.1 ("*Tu lucha es mi lucha*": *Mathematics for Movement Building*), Gloria Gallardo and Cathery Yeh examine disparities in pay that contributed to the Delano Grape Strike. Students consider the idea of a counternarrative and explore Larry Itliong's life to create a timeline of his journey as a labor organizer and his wages. As students compare the differential wages of farmworkers, they engage in decimal concepts and calculations and discuss how numbers were used to build solidarity among the different groups of farmworkers.

Figure 3.1. Five Mathematical Content Domains in Grades 3–5

Content Domain	Children Need More Opportunities to . . .
Whole-number concepts and operations	• Develop flexibility in reasoning with number and operation relationships. • Use subitizing activities across the grades to develop quantitative relationships. • Learn basic number combinations through sensemaking, not memorization. • Transition successfully and intentionally from additive to multiplicative thinking.
Fraction concepts and operations	• Use unit fractions as the building blocks for developing fraction knowledge. • See fractions as numbers whose magnitude can be represented on a number line. • Use real-world contexts for understanding fraction operations conceptually.

(Continued)

(Continued)

Content Domain	Children Need More Opportunities to . . .
Early algebraic concepts and reasoning	• Develop meaning for the equals sign as stating two expressions have the *same value*. • Discuss observations and intuitions about the properties and behaviors of operations. • Experience algebraic thinking across the mathematics curriculum.
Data concepts and statistical thinking	• Use data to describe the variability of phenomena in their world. • Create data displays to organize, analyze, and communicate information. • Use data distributions to answer questions and pose further questions.
Geometry and measurement concepts and spatial reasoning	• Develop spatial reasoning as an essential core of mathematical development. • Co-construct the meaning of attributes in two- and three-dimensional geometric shapes. • Discuss, understand, and quantify measurable attributes of shapes.

Source: Huinker (2020).

Mathematical Practices

How do I want students to engage in mathematical activity?

NCTM has long advocated for teaching mathematics through problem solving, using an inquiry-based approach. The inquiry process—problem posing, exploring, conjecturing, and problem solving—builds upon the mathematical strengths of students and engages them as *doers* of mathematics. TMSJ underscores the importance of students as active doers and sensemakers of mathematics, who author and generate mathematical strategies and share their mathematical insights, not as passive recipients of information. *Doing mathematics* involves engaging in the norms, routines, and habits that are central to the work of mathematicians.

> **TMSJ underscores the importance of students as active doers and sensemakers of mathematics, who author and generate mathematical strategies and share their mathematical insights, not as passive recipients of information.**

The Standards for Mathematical Practice (National Governors Association Center for Best Practices, Council of Chief State School Officers, 2010) are one such list of these habits of mind, and they are based on the "processes and proficiencies" stemming from the NCTM (1989) process standards and strands of mathematical proficiency from the National Research Council's (2001) report *Adding It Up*. A teacher who plans with these practices and processes in mind employs a broadened perspective of what is important within the mathematics that students need to learn. While the content standards center on what students are learning, the mathematics practices provide insight into the habits and dispositions students are developing through their engagement with mathematics.

Figure 3.2. Standards for Mathematical Practice

Standards for Mathematical Practice

1. Make sense of problems and persevere in solving them.
2. Reason abstractly and quantitatively.
3. Construct viable arguments and critique the reasoning of others.
4. Model with mathematics.
5. Use appropriate tools strategically.
6. Attend to precision.
7. Look for and make use of structure.
8. Look for and express regularity in repeated reasoning.

Source: Copyright 2010. National Governors Association Center for Best Practices and Council of Chief State School Officers. All rights reserved.

In Lesson 6.3 (*Modeling Library Funding*) by Hyunyi Jung and Megan Wickstrom, students are asked to use data about quality and numbers of books in different school libraries. Students use mathematics as a tool to make sense of their world and the experiences they have considering equitable sharing of resources. They engage in Mathematical Practice 4 (model with mathematics) as they notice the ways in which mathematics is used in their lives, specifically current concerns about library funding and resource distribution. The use of these mathematics concepts provides a window in which they can begin to explore, question, and justify concerns around inequitable access to resources across communities.

Social Justice Standards

How do I want students to engage in anti-bias, multicultural, and social justice issues to develop knowledge and skills to reduce prejudice and advocate for collective action?

Learning for Justice (formerly Teaching Tolerance) seeks to uphold the mission of the Southern Poverty Law Center to be a catalyst for justice, working in partnership with communities to dismantle white supremacy, and to advance the human rights of all people. Learning for Justice was created to prevent the growth of hate by reducing prejudice, and the Social Justice Standards (Learning for Justice, 2016) serve as a road map for anti-bias education at every grade level, including learning outcomes divided into four domains—Identity, Diversity, Justice, and Action—that recognize the skills and knowledge students need related to both prejudice reduction and collective action. Chapter 1 provides an overview of the Social Justice Standards with teacher and student actions for each of the four domains and examples of its application in classroom contexts. Figure 3.3 provides a list of the 20 anchor standards and Grades 3–5 learning outcomes. See Appendix E for a chart of the book's lessons correlated to the social justice standards.

Figure 3.3. Learning for Justice Social Justice Standards, Grades 3-5 Learning Outcomes

Anchor Standard	Grades 3-5 Learning Outcome
Identity 1	I know and like who I am and can talk about my family and myself and describe our various group identities.
Identity 2	I know about my family history and culture and about current and past contributions of people in my main identity groups.
Identity 3	I know that all my group identities are part of who I am, but none of them fully describes me and this is true for other people too.
Identity 4	I can feel good about my identity without making someone else feel badly about who they are.
Identity 5	I know my family and I do things the same as and different from other people and groups, and I know how to use what I learn from home, school, and other places that matter to me.
Diversity 6	I like knowing people who are like me and different from me, and I treat each person with respect.
Diversity 7	I have accurate, respectful words to describe how I am similar to and different from people who share my identities and those who have other identities.
Diversity 8	I want to know more about other people's lives and experiences, and I know how to ask questions respectfully and listen carefully and nonjudgmentally.
Diversity 9	I feel connected to other people and know how to talk, work, and play with others even when we are different or when we disagree.
Diversity 10	I know that the way groups of people are treated today, and the way they have been treated in the past, is a part of what makes them who they are.
Justice 11	I try and get to know people as individuals because I know it is unfair to think all people in a shared identity group are the same.
Justice 12	I know when people are treated unfairly, and I can give examples of prejudice words, pictures, and rules.
Justice 13	I know that words, behaviors, rules, and laws that treat people unfairly based on their group identities cause real harm.
Justice 14	I know that life is easier for some people and harder for others based on who they are and where they were born.
Justice 15	I know about the actions of people and groups who have worked throughout history to bring more justice and fairness to the world.
Action 16	I pay attention to how people (including myself) are treated, and I try to treat others how I like to be treated.
Action 17	I know it's important for me to stand up for myself and for others, and I know how to get help if I need ideas on how to do this.
Action 18	I know some ways to interfere if someone is being hurtful or unfair, and will do my part to show respect even if I disagree with someone's words or behavior.

Anchor Standard	Grades 3–5 Learning Outcome
Action 19	I will speak up or do something when I see unfairness, and I will not let others convince me to go along with injustice.
Action 20	I will work with my friends and family to make our school and community fair for everyone, and we will work hard and cooperate in order to achieve our goals.

Source: Reprinted with permission of Learning for Justice, a project of the Southern Poverty Law Center. www.learningforjustice.org

online resources ☞ Available for download at **resources.corwin/TMSJ-UpperElementary**

PAUSE AND REFLECT

Which elements of these learning outcomes are you already including in your lesson plans? Which will require further refinement?

ASSESSING PURPOSEFULLY

When teachers and students think about assessments, what often comes to mind are tests and quizzes and how these are used to assign grades and label students' process or achievement. This popular notion of assessment as evaluation neglects the important aspect of assessment as a method of gathering evidence of students' thinking to inform learning and instruction. According to Stefanakis (2002), the word *assess* comes from the Latin root *assidere*, which means *to sit beside*. In the educational context, "sitting beside" includes the process of observing and leveraging students' thinking to promote learning and improve teaching.

> In the educational context, "sitting beside" includes the process of observing and leveraging students' thinking to promote learning and improve teaching.

PAUSE AND REFLECT

What comes to mind as you reflect on the origin of the word *assess* and your own experience sitting next to a student making a breakthrough during problem solving, expressing pride in themselves and their work, or making principled decisions to convince and justify their stance in a mathematical task or for a justice-oriented cause?

Assessment Strategies

Which strategies can students and I use together to monitor students' understanding of mathematics (content and practices) and social justice issues?

Assessment is a central component of the teaching and learning process. It is both a part of teachers' reflections on instructional practices and students' autobiographical understanding of their own mathematics learning. Student learning is complex. Systems of assessments need the flexibility and capacity to capture such complexity in all its beauty and power.

Formative assessment tools are used during day-to-day interactions. The use of formative assessment gives you the opportunity to provide students with actionable feedback to support the discourse related to the mathematics content and the social justice issue. Emphasis is placed on the strengths students demonstrate during an assessment, rather than deficits. These assessments offer a starting point for educators to build on when designing classroom learning experiences by providing learners with feedback to move forward to activate students as knowledgeable contributors to classroom interactions. Figure 3.4, adapted from *A Fresh Look at Formative Assessment in Teaching Mathematics* (Silver & Mills, 2018), outlines some teacher supports of formative assessment strategies and connects them to discourse, which we'll discuss in the next section. We specifically identify strategies that support your actions to positively position your students and build their identity and sense of agency.

Figure 3.4. Assessment Strategies to Support Mathematics and Social Justice Discussions

Strategies for Supporting Formative Assessment	Connections Between Formative Assessment Strategy and Discourse
Providing feedback that moves learners forward	• Teachers are strategic about when to tell (e.g., when to show students what to do rather than letting them struggle through and figure it out). • Rather than rescuing their students when they are stuck, teachers have high expectations for how students should work with their groups. • Teachers sometimes explore incorrect answers. • Teachers monitor the room as their students are working. • Teachers use what they learned during monitoring to plan for productive discussions.
Activating students as the owners of their learning	• Teachers invite students to share their ideas. • Teachers position students as having the right to evaluate the reasonableness of one another's mathematical ideas. • Teachers position students as authors of mathematical ideas. • Teachers facilitate a growth mindset through discourse.
Activating students as resources for one another	• Teachers have students talk to one another in mathematics class. • Teachers make strategic use of group work. • Teachers use the think–pair–share strategy to give students an opportunity to think individually and to give all students opportunities to discuss their ideas. • Teachers provide students with accountable talk stems (Michaels et al., 2008).

Source: Silver & Mills, 2018.

 Available for download at **resources.corwin.com/TMSJ-UpperElementary**

Many assessment tools exist, and teachers must make decisions about which tools are best suited not only for the subject materials but also for their students. Figure 3.5 lists holistic assessment approaches to capture students' understanding of mathematics (content and practice), social justice standards, and the interplay of the two. The list is not exhaustive but is intended as a starting point to support your assessment practices related to SJMLs.

Figure 3.5. Holistic Approaches to Assess Students' Mathematics and Social Justice Understandings

Formative Assessment	Brief Description	Assessment of Mathematics Goals (Content and Practice)	Assessment of Social Justice Goals
Observation/ anecdotal notes	Informal and targeted observations of students engaged in mathematics learning. Teachers constantly gather evidence of student progress as they engage in the mathematics and social justice task.	Take time to record a few "look-fors" for the mathematics. For closed mathematics tasks, develop an answer key with anticipated answers. For open mathematics tasks, create several solution pathways or key features you expect to see in a solution.	Take time to record a few "look-fors" for the social justice issue. Anticipate multiple perspectives and various points of view. For the social justice issue, create several student responses (pro and con as appropriate) or key ideas or questions you expect to come up in discussion. Monitor students' emotions during small- and whole-group discussion.
Exit slips	Student responses, usually in written format, provide a quick, informal formative assessment. A teacher may pose a question or ask for a response to a comment, and allow students to express their thinking, understanding, or any further questions.	Write questions that support students in (a) sharing what they think that they learned, (b) sharing questions they have, or which (c) probe students' thinking and allow them the opportunity to demonstrate their understanding of content.	Write specific questions that address the lesson's social justice goal or standards. Develop question prompts to connect the mathematics and social justice issue.

(Continued)

(*Continued*)

Formative Assessment	Brief Description	Assessment of Mathematics Goals (Content and Practice)	Assessment of Social Justice Goals
Interviews	Brief, informal conversations between a teacher and a student or small group of students that provide a "deep dive" into student thinking and understanding. Teachers continuously monitor student progress and look for opportunities to support students.	Write questions that push and probe students' thinking. Include questions that address anticipated misconceptions and suggestions to build from students' strengths.	Write specific questions that address the lesson's social justice goal or standards. Develop question prompts to connect the mathematics and social justice issue.
Co-constructed assessments	Students and teachers work together in the creation of an assessment and its intended learning outcomes.	Together with students, identify mathematical learning goals and outcomes in ways that highlight and value student voice and experience.	Together with students, identify social justice standards, learning goals, and outcomes in ways that highlight and value student voice and experience.
Work samples/ portfolios	A sample of student work from a single activity, or a compilation of student work over time. An ongoing sample collection that provides evidence of student learning.	Take time to record a few "look-fors" for the mathematics. For closed mathematics tasks, develop an answer key with anticipated answers. For open mathematics tasks, create several solution pathways or key features you expect to see in a solution.	Take time to record a few "look-fors" for the social justice issue. Anticipate multiple perspectives and various points of view. For the social justice issue, create several student responses (pro and con as appropriate) or key ideas or questions you expect to come up in discussion.
Projects	Students actively engage in projects that provide real-world connections. Students and teachers reflect together on the authenticity of mathematics in real-world contexts.	Take time to record a few "look-fors" for the mathematics in project presentations.	Take time to record a few "look-fors" for the social justice issue. Anticipate multiple perspectives and various points of view.
Performances	Students produce a product that demonstrates their understanding and proficiency.	Take time to record a few "look-fors" for the mathematics in student performances.	Take time to record a few "look-fors" for the social justice issue. Anticipate multiple perspectives and various points of view.

 Available for download at **resources.corwin/TMSJ-UpperElementary**

Asset-Based Assessments

How might I consider what students do not yet understand in ways that view them as sensemakers rather than deficient?

In educational spaces, we are often trained to seek out and name the deficits of our students. We are asked to do this when we unpack test data and errors students make, when we are told to share with families where students need to develop in relation to grade- or age-level standards/norms, and when we make comparisons between students and their placement in trajectories of learning used in a one-size-fits-all context. Naming the strengths and unique insights and assets of our students is a far more powerful tool for planning purposes, as knowledge of students' understanding will allow us to make instructional decisions that build from where they are to progress toward the learning goals. Referred to as *funds of knowledge* (González et al., 2005), teachers recognize and draw upon the experiences of students in and outside of schools as members of a community. This recognition leverages knowledge connected to students' and families' social and cultural backgrounds and how each of us have intersections of social identities (e.g., race and ethnicity, language, gender, economic class, and disability). It also leverages the content knowledge students enter schools with. Such an orientation means the teacher is learning from the students as the students learn from the teacher. Both have expertise.

Recognizing the social, cultural, and academic expertise that students bring helps address a challenge many of us will experience as we implement SJMLs. We have felt a tension in our work teaching mathematics for social justice—how to know what is the "right" answer to a social dilemma or injustice. And further, what is our role in communicating to students the "correctness" of their personal conclusion? Drawing on our knowledge of teaching mathematics, we have learned to value many valid pathways to a solution, and even possibly many valid solutions to mathematics tasks. We have learned that we can rely on our experiences teaching mathematics in which we draw out student ideas in order to devise parallel strategies for the discussion of student ideas about a social context or injustice. We ask students to support their conclusions with evidence and to express curiosity about the contributions and experiences of others in an open-minded way. We ask students, and ourselves, to listen closely to student ideas and perspectives and to respond to diversity by building respect, understanding, connections, and empathy. Or we ask students to state the moral or ethical grounding (or assumption) that leads to the conclusion they make. Finally, as we rely on the Social Justice Standards Grades 3–5 learning outcomes to identify learning goals for the lesson, rather than aiming for students to have the *right* answer to a social injustice, we can ensure that the classroom experiences and conversations serve the social justice goal we set for the lesson.

When teachers focus too strongly on student mathematical errors and misconceptions, students become wary of sharing ideas in the classroom. The same principle holds true if teachers silence ideas related to social justice issues; this would be counter to the aims. Teachers must develop a classroom culture that values students sharing ideas—mathematical or critical, even when they are in "rough-draft" form (Jansen et al., 2017)—and then ask students to shape, debate, and refine those ideas as a part of the lesson. It is your role as the teacher to have some say in identifying where student ideas conflict with experts in the area, or even when there is disagreement among people who are deeply involved in the topic. Rather than viewing your students as incorrect or ignorant, however, we suggest examining students' current conceptions or funds of knowledge to understand what may lead them to the conclusions they reach. Using effective questioning and discussion strategies can help in drawing out students' thinking and their justification for such thinking in productive ways. This places emphasis on students sharing their knowledge openly, providing opportunities for teachers to examine their current understanding and where it comes from in their rich history.

TEACHING EQUITABLY

Students who have the opportunity to critically examine issues of social injustice become further-informed citizens and active participants in a diverse society. However, this only happens with intentional instruction. Students experience the world in vastly different ways. When topics and voices are expressed that do not reflect the dominant or majority view, many students have the opportunity to feel recognized, opening the potential to develop a sense of empowerment. These voices, however, can easily be shrouded by the dominant view—that which is assumed to be normal in the local setting. TMSJ teachers can draw upon equitable teaching practices to prevent this from happening.

The NCTM publications *Principles to Actions* (2014), *The Impact of Identity in K–8 Mathematics: Rethinking Equity-Based Practices* (Aguirre et al., 2013), and *Toward a Framework for Linking Equitable Teaching with the Standards for Mathematical Practice* (Bartell et al., 2017) identify research-based instructional practices that exemplify important qualities of mathematics instruction. *Principles to Actions* provides a set of eight mathematics teaching practices. The *Impact of Identity* names five equity-based instructional practices that support teachers in the use of these eight mathematics teaching practices. *Toward a Framework for Linking Equitable Teaching* identifies additional equitable mathematics teaching practices to support teachers in the use of the eight mathematics teaching practices. Figure 3.6 outlines these practices side by side for review.

Figure 3.6. Mathematics Teaching Practices (NCTM, 2014) and Equity-Based Mathematics Teaching Practices (Aguirre et al., 2013; Bartell et al., 2017)

Mathematics Teaching Practices	Equity-Based Teaching Practices
1. Establish mathematical goals to focus learning.	1. Go deep with mathematics.
2. Implement tasks that promote reasoning and problem solving.	2. Leverage multiple mathematical competencies.
3. Use and connect mathematics representations.	3. Affirm mathematics learners' identities.
4. Facilitate meaningful mathematics discourse.	4. Challenge spaces of marginality.
5. Pose purposeful questions.	5. Attend explicitly to race and culture.
6. Build procedural fluency from conceptual understanding.	6. Draw on multiple resources of knowledge (math, culture, language, family, community).
7. Support productive struggle in mathematics.	7. Support development of a sociopolitical disposition.
8. Elicit and use evidence of student thinking.	

We hope the lessons in this book provide an opportunity for you to build important connections between your own teaching and these equity-based mathematics teaching practices. Filiberto Barajas-Lopez and Gregory Larnell (2019) remind us, however, of the dangers of linking equity-based teaching practices to standards (e.g., the Standards for Mathematical Practice) as this cooperates tacitly with the Common Core State Standards for Mathematics in ways that might reproduce the inequalities we are intending to eliminate and limits potential to take up more robust understanding about culture, race, and families and communities. Thus, as you take up these equitable teaching practices, you should be challenging notions about the nature of learning, culture, and race as well as accounting for the emergent voices, experiences, and perspectives of students' families and communities.

Resources in this book help you center and create authentic experiences for students while also generating mathematics- and justice-based questions that probe particular issues and situations relative to students' lives and society. Activities in this book are designed to encourage student-to-student interactions that move beyond the classroom to actionable change in their lives, their communities, and possibly even for the broader society.

As you take up these equitable teaching practices, you should also challenge notions about the nature of learning, culture, and race as well as account for the emergent voices, experiences, and perspectives of students' families and communities.

PAUSE AND REFLECT

Ask yourself the following:

- During my instructional planning, what are some of the equity-based practices that I readily attend to and consider in lesson and classroom design?

- What are some of the equity-based practices that I need to focus on more carefully?

NAVIGATING DISCOURSE

Many of the lessons in this book utilize instructional strategies that encourage the development of individual students' ideas and considerations of others' thoughts. Included in these strategies is the think–pair–share instructional routine. For example, in Lesson 7.1 (*Water Is Our Right, Water Is Our Responsibility*), think–pair–share is used to support student discussion at the beginning of the lesson of how much water students think is consumed in one day. This routine is used again to support students in processing new information and discussing if they want to revise their thinking. Other lessons, such as Lesson 7.2 (*Upper Elementary Math to Explore People Represented in Our World and Community*) and Lesson 6.3 (*Modeling Library Funding*), have students use "notice and wonder" protocols to share their thoughts, insights, and beliefs about data, pictures, or video content. In this section, we discuss a few foundational principles of the role of the teacher in the discourse-rich TMSJ classroom—specifically, strategies for increasing engagement, valuing student voice, and organizing for difficult conversations.

Discourse to Keep Students Engaged

During my instructional planning, how can I design opportunities and structures for meaningful discourse that is related to the lesson goals and engages my students?

How teachers position students to speak and be heard in the mathematics classroom is one of their most important responsibilities.

An important part of facilitating a SJML is creating an environment of safe and open discourse among the students and with the teacher. This discourse often extends to other teachers, school leaders, families, and community stakeholders. We know that classroom discourse does not happen by accident, nor is it an automatic component of particular lessons. How teachers position students to speak and be heard in the mathematics classroom is one of their most important responsibilities. The quality of discourse also depends on the classroom environment and culture that has been previously established by the teacher.

As you plan to facilitate a SJML, the role of discourse is critical to keeping students engaged in the lesson. Here are three questions to consider when planning to integrate a SJML into your classroom. Your responses to these questions should tell you how engaged your students might be with lessons that involve a social context, especially one that might be sensitive.

1. How are you being attentive to student concerns or questions about their school or community (or world)?

2. How are you staying connected with students' communities and realities?

3. Why would students see you as a co-conspirator (Love, 2019) for them and other socially and historically marginalized people in their community?

You may also want to consider the *5 Practices for Orchestrating Productive Mathematics Discussions* (Smith & Stein, 2018) as a resource when planning classroom discourse. This will help ensure that your plans anticipate students' responses as you monitor their work and select, sequence, and connect students' responses in meaningful and equitable ways to highlight the mathematical and social justice goals in each lesson. Figure 3.7 lists the five practices; we have added other ideas for you to consider as you plan to use these strategies when integrating issues of social injustice into the mathematics classroom.

Figure 3.7. Practices for Orchestrating Productive Discussions When Planning the SJML

5 Practices – Mathematics Discourse	Social Justice Discourse
Anticipating likely student responses to challenging mathematical tasks and questions to ask students who produce them	**Anticipating** likely student points of view and asking questions that help students identify specific points of agreement and disagreement
Monitoring students' actual responses to the tasks (while students work on the tasks in pairs or small groups)	**Monitoring** students' ideas or positions to keep them focused on and grounded in the topic
Selecting particular students to present their mathematical work during the whole-class discussion	**Selecting** particular students to share their outcomes or decisions with support so that diverse perspectives and voices are represented
Sequencing student responses that will be displayed in a specific order	**Sequencing** student responses so that different perspectives and opposing points are shared and valued
Connecting different students' responses and connecting the responses to key mathematical ideas	**Connecting** different students' responses and perspectives to help students identify places where compromise might be possible, and connecting to key social justice standards

Source: Adapted from Smith and Stein (2018).

 Available for download at **resources.corwin/TMSJ-UpperElementary**

Valuing Voice

How can I ensure that mathematical discourse—students actively speaking and being heard—is a prominent and equitable part of each student's experience of our classroom culture?

Valuing each and every student's voice is a hallmark of a classroom grounded in TMSJ and equitable teaching practices. As a teacher, your actions and voice set the tone for establishing a classroom culture that nurtures acceptance, respect, and tolerance. The manner in which you speak and listen to your students in whole-class, small-group, or student-to-teacher settings establishes the model students will follow. Figure 3.8 suggests ways that the teacher may promote or discourage discourse during a SJML.

Students who feel a sense of respect, acceptance, and value from their teacher are more likely to behave similarly toward their peers. Modeling equitable speaking and listening skills with students will help establish a classroom culture that makes room for discourse about controversial topics as part of lessons. You can be intentional in highlighting when students demonstrate speaking and listening skills that are productive to fostering respectful discourse.

Figure 3.8. Teachers' Actions to Facilitate Discourse During a SJML

	Promotes Discourse	Discourages Discourse
Verbal patterns	Facilitates and referees debate between students in a way that promotes student identity	Allows students to use language and tone that reduce the voice and experiences of other students
	Uses questioning probes and revoicing strategies that ensure equity and student ownership of thought	Moves from one mathematical or social discussion to the next without ensuring student voice was understood by others
	Uses open questioning to facilitate discourse, which allows students to create and maintain ownership and voice	Uses closed questions, which often include one-word responses that funnel students into a preplanned thought pattern
	Uses good "wait time" to promote discourse among students	Answers student questions quickly in group and classroom discussions without seeking student voice
Nonverbal patterns	Uses tasks with multiple entry points, pathways to a solution, or differing solutions	Uses tasks with one solution or one potential pathway to a solution often found in the directions of an activity or scaffolded exercises
	Chooses and uses student thought, perspective, and experiences to ensure connection to social justice and mathematics	Does not seek equitable participation by calling on students and/or purposefully disregards students' solutions, voice, and/or perspective

 Available for download at **resources.corwin/TMSJ-UpperElementary**

In addition to highlighting moments when a student exemplifies a strategy that contributes to a classroom climate that respects student ideas, these actions that promote respectful discourse can be identified and named as valuable strategies. By posting these in the room and referring back to them regularly, you can ensure they become a part of your classroom norms. Dr. Julia Maria Aguirre, of the University of Washington-Tacoma, and her colleagues use the following discussion guidelines to establish classroom expectations for respectful discourse (Aguirre et al., 2018):

- Prepare to feel some discomfort.

- Listen respectfully.

- Share the time.

- Be mindful of not just the intent of your words, but their impact as well.

- If you feel the need to challenge something said, be sure you are challenging the ideas, and not the people who shared them.

SUPPORTING ALL STUDENTS—CULTURALLY RESPONSIVE AND INCLUSIVE MATHEMATICS CLASSROOMS

Consider how your students would answer the following questions:

- During lessons, how are your ideas valued?

- How does learning mathematics connect to joy, wonder, and relevance for you?

Students' responses are a gauge of the sort of mathematics learning environment created. They reveal the extent to which students feel seen, valued, and included in the learning of mathematics and the degree to which meaningful, rigorous, and wonder-filled mathematics lessons are made available to them. All students have the capacity to bring curiosity, joy, and wonder to their engagement with mathematical ideas, whether in the classroom or in their lives outside of school.

The lessons in this book were prepared with a diverse student population in mind. For each lesson, the design, task structure, mathematical and social justice content, and teaching practices were intentionally selected to support students' exploration of mathematical concepts and communication of their ideas, regardless of areas of mathematical and cultural strengths and backgrounds. We build from an understanding that classrooms have not always been designed as ones in which all students—particularly students of Color, emergent bilinguals, and students with disabilities—were positioned as valuable resources; their brilliance is often hidden and not yet leveraged, due to educational barriers, to promote individual and collective success. We use the term emergent bilinguals in this work, which we prefer over English Language Learners. Although the term English Language Learners is widely used, it fails to highlight one of the most important aspects of the majority of students learning English in schools: they are bilingual. In fact, some of the students who are speaking a language other than English at home already speak two or more languages by the time they learn English in school. Students' bilingual and bicultural identities should be leveraged as an asset.

Mathematics lessons to address social (in)justice require us to dismantle systems and categories within our own classrooms that demarcate and rank our students along lines of power and privilege. We bring ideas of culturally relevant pedagogy (CRP; Ladson-Billings, 2009) and Universal Design for Learning (UDL; CAST,

> All students have the capacity to bring curiosity, joy, and wonder to their engagement with mathematical ideas, whether in the classroom or in their lives outside of school.

2018) as a means to build justice-oriented pedagogies that attend to intersecting markers of difference (e.g., disability, class, gender, race and ethnicity, language). An important aspect of this work requires us to create opportunities to have students' identities (their experiences, communication practices, and communities) reflected within their learning of mathematics. UDL is a framework that focuses on eliminating educational barriers that result in inequitable learning opportunities and outcomes so that the learning experiences engage learners, activate thinking, and scaffold deep understanding. UDL is organized through three broad principles, which are aligned with three networks in the brain that involve the learning processes (see Figure 3.9):

1. Multiple means for engagement of students (corresponding to the *affective* network—the *way* of learning)

2. Multiple means of representation of information to students (corresponding to the *recognition* network—the *what* of learning)

3. Multiple means for *action and expression* by students (corresponding to the strategic network—the *how* of learning)

The framework embraces lesson development that works for everyone—not a single, one-size-fits-all solution but rather flexible approaches that honor student diversity. Each and every student has unique strengths, skills, and interests and a different level of comfort with the mathematics and social justice content.

To eliminate barriers that result in inequitable learning opportunities and outcomes, for each part of a lesson, identify the primary goal: What do you really want students to know, do, or care about for that part of the lesson? Then, reflect on the different pathways or options students have to achieve that goal. That flexibility helps reduce barriers and increases meaningful learning opportunities. Ask yourself the following questions:

- Where might there be a barrier to students achieving the goal in this lesson?

- What is one tool, resource, or strategy I can include in my lesson to help reduce this barrier so that students can achieve the learning goal?

Below are some strategies that cross-pollinate CRP and UDL to leverage the brilliance of and diversity in all students (Yeh & Chao, 2019).

- Position learners as experts in and about their learning, their schools, and communities.

- Expand ideas about what doing mathematics entails and what counts as mathematical competence beyond speed and accuracy. Publicly draw attention to competencies aligned with the Common Core Standards for Mathematical Practice.

Figure 3.9. Universal Design for Learning Guidelines

The Universal Design for Learning Guidelines

CAST | Until learning has no limits

Provide multiple means of **Engagement**	Provide multiple means of **Representation**	Provide multiple means of **Action & Expression**
Affective Networks The "WHY" of Learning	Recognition Networks The "WHAT" of Learning	Strategic Networks The "HOW" of Learning

Access

Provide options for **Recruiting Interest** (7)	Provide options for **Perception** (1)	Provide options for **Physical Action** (4)
• Optimize individual choice and autonomy (7.1) • Optimize relevance, value, and authenticity (7.2) • Minimize threats and distractions (7.3)	• Offer ways of customizing the display of information (1.1) • Offer alternatives for auditory information (1.2) • Offer alternatives for visual information (1.3)	• Vary the methods for response and navigation (4.1) • Optimize access to tools and assistive technologies (4.2)

Build

Provide options for **Sustaining Effort & Persistence** (8)	Provide options for **Language & Symbols** (2)	Provide options for **Expression & Communication** (5)
• Heighten salience of goals and objectives (8.1) • Vary demands and resources to optimize challenge (8.2) • Foster collaboration and community (8.3) • Increase mastery-oriented feedback (8.4)	• Clarify vocabulary and symbols (2.1) • Clarify syntax and structure (2.2) • Support decoding of text, mathematical notation, and symbols (2.3) • Promote understanding across languages (2.4) • Illustrate through multiple media (2.5)	• Use multiple media for communication (5.1) • Use multiple tools for construction and composition (5.2) • Build fluencies with graduated levels of support for practice and performance (5.3)

Internalize

Provide options for **Self Regulation** (9)	Provide options for **Comprehension** (3)	Provide options for **Executive Functions** (6)
• Promote expectations and beliefs that optimize motivation (9.1) • Facilitate personal coping skills and strategies (9.2) • Develop self-assessment and reflection (9.3)	• Activate or supply background knowledge (3.1) • Highlight patterns, critical features, big ideas, and relationships (3.2) • Guide information processing and visualization (3.3) • Maximize transfer and generalization (3.4)	• Guide appropriate goal-setting (6.1) • Support planning and strategy development (6.2) • Facilitate managing information and resources (6.3) • Enhance capacity for monitoring progress (6.4)

Goal

Expert learners who are...

Purposeful & Motivated	Resourceful & Knowledgeable	Strategic & Goal-Directed

| © CAST, Inc. 2018 | Suggested Citation: CAST (2018). Universal design for learning guidelines version 2.2 [graphic organizer]. Wakefield, MA: Author.

Source: CAST (2018) Universal design for learning guidelines version 2.2 [graphic organizer], Wakefield, MA. Used with permission. All rights reserved.

- Value all student voices, regardless of time and modality. Mathematics communication is not just talking and writing. Allow students to express their ideas using their choice of modalities, with some sharing their thoughts out loud and others using drawings/illustrations, a podcast, a video, or even a comic strip.

- Allow students access to resources (assistive technologies such as speech-to-text, alternative keyboards), materials, and opportunities to use their first languages as part of their cultural repertoires to create and share ideas and final products.

- Foster a culture emphasizing genuine collaboration while deemphasizing competition. Select activities for interdependence, creating opportunities for students to play with the mathematics on their own and in groups with flexible composition (e.g., gender, ability, language) and roles (e.g., facilitator, reporter, notetaker).

All students are served well in a discourse-rich, multidimensional classroom in which all students have some mathematical strengths but none are viewed to be strong in everything. When given rich tasks, students need the mathematical (and social and cultural) assets and strengths of all of their peers in order to be successful with the task. Of course, when a required accommodation has been identified, it is still appropriate to use when engaging in TMSJ.

Dealing With Emotional or Sensitive Conversations

What sensitive issues or varying views, feelings, and perspectives should I be prepared to acknowledge, support, and validate?

As you plan for your lesson, we encourage you to carefully consider the classroom environment and culture when selecting the social justice issue. Earlier, we discussed establishing a classroom environment in which ideas are discussed, not people, and that all of us learn together. Rather than focusing on how a fellow student is wrong, seek to understand the reasoning that brought them to their conclusion.

In addition to establishing a classroom climate focused on ideas and thinking, you will need to plan for a contentious topic, for how you will sustain your classroom as a community through emotional discussions, and for students to communicate, actively listen, reflect, and work through ideas to understand diverse perspectives. You need to be ready to answer students' questions and help them understand the issue(s), even as you might be struggling to understand them yourself. Each of us brings different knowledge and a different level of comfort to leading discussions on contentious topics. In *Courageous Conversations About Race: A Field Guide for Achieving Equity in Schools*, Glenn Singleton and Curtis Linton (2005) suggest four agreements for participants engaging in such courageous conversations:

1. Stay engaged.

2. Experience discomfort.

3. Speak your truth.

4. Expect and accept nonclosure.

Further, to sustain a community through emotional or tough discussions, it is imperative that both the teacher and students work to build a community of active listeners "who feel safe displaying their full identities but also who receive the identities and cultural experiences of others" (Howell, 2020). This does not mean you have to be in agreement with what is being said, but rather includes moving beyond selective hearing to listening for details and modeling respectful curiosity. We recommend that you learn more about some potential activities through Facing History and Ourselves' "Current Events Teacher Checklist" (available at bit.ly/2kvsdTK).

In many lessons that examine social injustices, the names and terms we use for the people involved can potentially serve to further marginalize them, in terms of both their own sense of identity and how others perceive them. Terms matter

> You need to be ready to answer students' questions and help them understand the issue(s), even as you might be struggling to understand them yourself.

for students (people) deprived of representation and can open a door toward the development of a more positive identity. Further, a definition can point to a community, making a person feel less alone. Identity is important.

We view the mathematics classroom as a place where students learn to exchange ideas, listen respectfully, and validate and share differing perspectives to ideas. Our aim is that all students can experience this discourse-rich classroom without fear or intimidation. Many teachers have refined the development of this classroom culture for discussions of mathematics but may feel a small amount of intimidation to bring issues of social injustice into that same climate. Luckily, engaging in this work does not require us to know it all but to model the humility, respectful curiosity, and active listening needed to work and learn across differences. "All communities are full of differences, and those differences can clash in ways that will break us if we let them. We have to prepare our students—and ourselves—to communicate, question, and work our way through" (Howell, 2020). Below is a list of strategies to prepare you and your students to communicate, question, and work through tough conversations by making active listening the norm:

- *Make sharing the classroom norm:* Encourage students to share through modeling that vulnerability by sharing your own stories.

- *Practice listening with intention:* Have students write down or rephrase what their classmates are saying to ensure each student's whole message and not just partial messages are heard.

- *Model respectful curiosity:* Encourage students to ask follow-up and clarifying questions to promote meaningful discourse.

We know that many topics will be very meaningful to our students because they reflect issues students are experiencing right now, deconstruct stereotypes that impact them and their classmates, and engage students in learning about the untold stories, contributions, and histories that allow them to expand perspectives and see themselves and each other as part of the narrative of the United States. Engaging students on topics leading to diverse perspectives and opinions provides them the opportunity to develop critical thinking skills, build interpersonal communication skills, empathize with the conditions and circumstances of others, honor differences even when not everyone agrees, and develop a sense of civic responsibility. We encourage you to rely on the lessons in this book and the teaching practices above to plan for these meaningful, difficult, and important conversations.

CONCLUSION

If this is your first foray into implementing SJMLs, you may have some trepidation over how things will turn out in your classroom. We report on the experiences of several teachers and the lesson authors in Chapters 5–7. While you cannot predict every detail of how a lesson will unfold in your classroom, we've all experienced the power of pausing to consider how students might respond to a prompt or a task, how to better structure an activity to ensure equitable participation and discussion, how to engage responses to ideas rather than people, or how to justify a response through previous knowledge and facts.

Our emphasis is that you already possess many of the skills to implement a SJML effectively. You might learn some additional strategies to support student-to-student discussion about sensitive or emotional topics, or you might refine your formative assessment strategies to focus on your social justice goals in addition to your mathematical goals. Ultimately, TMSJ presses you to be a more effective teacher of children. During mathematics time, mathematics doesn't have to be the only thing foregrounded. Students' interests and concerns about their lives, their community, and their world can provide you with rich context to develop informed and active citizens who are able to participate in and shape a diverse society.

Students' interests and concerns about their lives, their community, and their world can provide you with rich context to develop informed and active citizens who are able to participate in and shape a diverse society.

REFLECTION AND ACTION

Take a moment to pause and reflect on your instructional practices. What is your level of comfort with establishing mathematics and social justice goals? What assessment techniques might you need to adjust or learn? How ready are you to facilitate discussions, not just about mathematics but about topics that might lead to emotional or divisive responses?

One of the following steps can help you begin implementing one of the lessons in this book:

1. Identify one Social Justice Standard (see Appendix D) that resonates with you. Select one discourse strategy that can help activate that standard to integrate into your class.

2. Identify one Mathematics Content Domain (see Appendix C) that resonates with you. Select one discourse strategy that can help activate that mathematics domain to integrate into your class.

3. Consider interdisciplinary connections with Social Studies or English Language Arts that might engage students in critical analysis of important issues.

TEACHING THE SOCIAL JUSTICE MATHEMATICS LESSON

CHAPTER
4

Each and every child should develop deep mathematical understanding as confident and capable learners; understand and critique the world through mathematics; and experience the wonder, joy, and beauty of mathematics.
(NCTM, 2020, p. 11)

Now that we have provided some background on the purposes, strategies, and pedagogical tools for teaching mathematics for social justice (TMSJ), you can begin planning a social justice mathematics lesson (SJML) that may help you achieve the aims set in this opening quote from NCTM as the purpose of elementary mathematics. You will find that the lessons in this book were asked to follow the Social Justice Mathematics Lesson Plan format (see Figure 2.8). However, we had a variety of format types submitted. You will notice that while there is a common flow to the lessons, they have varying scopes and sizes. Most lessons begin with an opportunity for students to learn more about the social justice issue and its contexts. Students are encouraged to wonder and ask questions, providing opportunities for them to create meaningful and personal connections to the issue. This begins an authentic, sustained inquiry into the task. Although most of the lessons in this book are designed to last for 2–5 days, some could be extended even further. Each lesson closes with students taking action, a form of a public product, and includes ways for teachers to communicate with various stakeholders. We hope that you will look at these lessons, review them to understand how they were implemented, and envision what they could look like in your own context, making any necessary modifications to meet the needs of the students in your classroom.

PLANNING TO IMPLEMENT A SJML

The elements of the SJMLs we have included in this book were designed to help teachers think through different, important aspects of mathematics lessons designed to explore, understand, and respond to social injustices. The lessons in this book were not meant to be picked up and followed exactly; these lessons were meant to be examples of what is possible in upper elementary classrooms. Like any curricular resource, you will likely read through the lesson and see parts of the lesson that speak to you that you can use as is and others that you may have to modify to reflect your context.

The lessons in this book are designed to engage students in meaningful mathematics to more fully understand issues of social injustice and be able to act. As with any mathematics lesson, we encourage you to be ready to support multiple ways of thinking about solutions, both in terms of mathematics and thinking about responding to the social justice issue at the center of the lesson.

We encourage you to structure your lessons in a way that helps you enact the equitable mathematics practices introduced and discussed in previous chapters. These practices are foundational to TMSJ and help teachers establish a classroom culture and environment that is ready to take on the challenge of integrating social justice lessons in the mathematics classroom. We encourage you to revisit them when thinking through your lessons.

How a teacher implements a lesson is not a linear, predictable, or systematic process, as the structure of the SJMLs we present in this book may suggest. Many excellent SJMLs arise in classrooms in which teachers listen to and are involved with students and have established a culture of facilitating social justice goals that integrate mathematical standards. Reflections on these experiences and recommendations from the lesson authors are provided in Chapter 8.

COMMON STRUCTURES FOR ALL SJMLS

As you prepare to implement any of the SJMLs provided in this book, we recommend that you start by carefully reading through the materials provided by the lesson author(s). Most lessons require preparation of materials for your particular context. Each SJML begins with a brief description and overview of the lesson describing the social justice topic that the lesson investigates.

Social Justice Topic and Brief Description

In this section, the lesson author(s) introduce to you what the social justice topic is and the connection and significance of the topic to them and the students in their classroom. The topics in this book are grounded in various issues important and relevant in upper elementary education. They are issues that could arise from questions or concerns from students, provoke them, or connect to their prior knowledge and understandings, allowing for authentic and challenging learning. The social injustices explored cover ability, class, environmental justice, economics, health, opportunity, race and ethnicity, and rights and activism. These contexts can help students to learn about themselves, to learn about other students in their classrooms and communities, to learn about the world around them, and to learn mathematics. These lessons also provide opportunities to observe patterns, to critique information, to learn to ask questions, and to reflect.

As noted, the contexts of the lessons here may be less authentic to the students in your classroom given that they were written by other people. The local, authentic context can serve as a powerful way to increase student engagement and motivation to learn mathematics, understand a social injustice, and plan and carry out collective action. Curriculum can and should serve as "windows, mirrors, and sliding glass doors" (Bishop, 1990, p. ix) for students to experience and understand the world. In other words, our teaching should reflect students' lives and experiences, give them insight into others' perspectives, and allow them to enter into new spaces and understandings.

Mathematics Content and Practice Standards and Social Justice Standards

Each of the lessons identifies and provides an opportunity to support, challenge, and expand students' knowledge of three goals:

1. *Mathematics Content*—what we want students to know and be able to do

2. *Mathematics Practices*—how we want students to show what they know and can do (National Governors Association Center for Best Practices, Council of Chief State School Officers, 2010)

3. *Social Justice Standards*—how we want students to demonstrate their understanding of and response to an issue (Learning for Justice, 2016)

We especially encourage you to return to the Social Justice Standards, and specifically the Grades 3–5 Outcomes (Learning for Justice, 2016), to consider how you might refine the learning goals you have for the students in your classroom. As you have a better understanding of the social justice goal(s), you also have a foundation from which to decide what courses of action to pursue.

Deep and Rich Mathematics

Although we asked the lesson authors to state the mathematics concepts that their lesson was aligned to, we also asked them to describe more fully the mathematics in the lesson and the ways in which the lesson could empower students mathematically. It is in this section that teachers may also see the connections to their grade level and be able to envision modifications to meet the needs of their particular students.

Materials and Resources

Like most curricular resources, the lessons in this book list materials you need for the lesson, including handouts, children's books, mathematics manipulatives, and other kinds of teacher materials and resources. Printable versions of handouts and other kinds of resources are available for download at resources.corwin.com/TMSJ-UpperElementary.

Lesson Plans for Multiple Days

Daily plans within each lesson include detailed procedures that walk you through how the author(s) implemented each lesson. We recommend that you review the lesson plans once to identify the social justice and mathematics goals and then map the overall structure of the lesson onto the context of your classroom. Determine how you will assess student attainment of these goals and make any needed modifications to best match the interests and experiences of your students and match your instructional goals. You may want to revise the lesson's context to localize it to your school or community setting; for example, Lesson 7.3 (*Single-Use Plastics*) is written using information for California, so you might consider using the appropriate information for your city or state.

The SJMLs selected for this book provide opportunities to build on the individual, family, cultural, and social knowledge that students bring to your class and to challenge spaces of marginality, specifically centering on students' experiences and knowledge as legitimate intellectual spaces for investigation of mathematical ideas. As you read through the lessons, be intentional about thinking about the students in your classroom and their individual and collective strengths and knowledges, and how you might leverage those to strengthen the learning experience.

Introducing the Lesson

TMSJ requires teachers to be attentive to developing their students' social, cultural, and mathematical identities. Many students' attitudes and perspectives are largely shaped by their past experiences in mathematics classes and their own lens through which they view the world. For this reason, many of the lessons introduce students to the social context at the beginning. While there are numerous ways to introduce the social context, it is important that this introduction sparks an interest in students to further explore, which will create an intellectual and possibly social or emotional need that will drive student learning. We highlight three options that can be effective to begin a SJML: storytelling, the use of articles or videos, and mathematics.

One effective strategy for introducing a SJML is the use of storytelling. Delgado (1990) calls this pedagogical practice a powerful opportunity for marginalized groups to draw attention to their own experiences and provide insights to others through their own lens. Two examples in this book that use storytelling to introduce a social issue are Lesson 7.2 (*Upper Elementary Mathematics to Explore People Represented in Our World and Community*) and Lesson 6.1 (*"Tu lucha es mi lucha": Mathematics for Movement*). In Lesson 7.2, students read excerpts of the book *If the World Were a Village* and watch a video to explore the characteristics of the village. They consider what they notice and wonder about the data presented. In Lesson 6.1, students begin by telling stories through pictures, drawing their image of a farmworker. In these lessons and many others, the authors offer a way for students to become more familiar with a social concern through the lens of someone involved in or knowledgeable about a situation.

Articles or videos are another method that provide students the opportunity to connect to the social context of the lesson by drawing upon students' social, cultural, and academic funds of knowledge (González et al., 2001). For example, they may recall and build upon their knowledge of food sharing at family meals in Lesson 5.4 (*Family Story Problems*) or connect to a video provided by the teacher at the beginning of the lesson, such as in Lesson 5.2 (*Playground Prejudice*). A central question helps students focus their reading or viewing, and the lesson(s) offer additional questions for a teacher to facilitate a discussion. In Lesson 7.3 (*Single-Use Plastics*), students view a brief video as an introduction to the social injustice. After the video, they are asked what they notice and wonder and are posed mathematics questions related to the video.

Another route that some authors have taken to introduce a lesson is to focus on the mathematics of the lesson first, allowing the mathematics to unearth or highlight the social concern. Gloria Ladson-Billings (2019) shared a story of a lesson taught by Bill Tate, currently President of Louisiana State University in Baton Rouge. He offered students $\frac{1}{114}$ and $\frac{1}{4310}$ for comparison mathematically with no context. As students began to see how different the two fractions were, they were introduced to the context of these fractions. In 1840 in Massachusetts, $\frac{1}{114}$ of slaves were declared insane; however, $\frac{1}{4310}$ of slaves in Georgia were declared insane (Deutsch, 1944). Students were then asked to speculate why they believed the fractions were so different. Before the Civil War, several Southern states had laws to restrict or deny mental health care for Black people, such as a statue in Virginia that "decreed that 'no insane slave should be received or retained in either (the Eastern or Western Lunatic) Asylum so as to exclude any white person residing in the state'" (p. 480). In Dr. Tate's lesson, he drew in the learner's interest through mathematics, opening the opportunity to consider the context of a social injustice and possibly further pursue the mathematics. In Lesson 5.3 (*Who Appears in the Billboards?*), students begin by collecting data (snapshots of billboards) and then work to categorize and identify the social identities in the billboards. Students use their data collection and sorting (mathematics) to uncover a social concern rather than the social concern being examined at the onset. In both situations, students' engagement with and understanding of mathematics plays a role in their understanding of a social injustice.

Facilitating the Exploration

The primary roles of the teacher while students work are to facilitate student discourse, question to probe and push students' thinking, and use formative assessment strategies to learn students' level of understanding of both the mathematics and social justice goals. Chapter 3 offers several strategies to facilitate student-to-student discourse. The lessons in this book include deep mathematical and social justice–related questions to support you in this work. The questions appear throughout each part of the lesson materials.

During the opening of a lesson, many students may be reluctant to openly share their experiences around bullying from other classmates, sexism, racism, and so on out of fear of how they may be perceived by their peers. Similarly, students are less apt to share their mathematical thoughts without opportunities for self-reflection and peer checking. Using the deep mathematical and social justice questions in small groups supports students in grappling with thoughts and how their responses might be perceived by their peers at a smaller level. In addition, teachers may purposefully group students in ways that provide support for student sharing of ideas. During the exploration portion of a SJML, you should monitor student progress both mathematically and socially. Consider

- Purposefully monitoring student explorations and facilitating these pathways toward meaningful mathematical and social discoveries,

- Completing brief notes on students' mathematical pathways to resolving the social injustice, and

- Looking for students who have stories to tell that relate to the SJML and providing them with the opportunity to share with others outside of their small group.

In Lesson 6.6 (*Your Action Saves Lives: COVID-19 and Systems Thinking*), students investigate a number of statistics related to health care. Factors that may be explored in the data include the number of breaths per minute that 6- to 12-year-olds take, the average cost of visiting an emergency room, the number of COVID cases, and the number of children and teens in Los Angeles who have asthma. The lesson authors suggest allowing time for students to explore the data and consider where it might relate to a health care systems map recently created by the class. During this time, teachers encourage students to ask one another questions such as these: *How much is that? How is that value significant? Why does this data matter? How does this data impact our understanding of health care justice?* As students explore, teachers may consider taking notes on students' thoughts for further discussion in the whole group using a monitoring sheet.

Closing the Lesson

As you help students to identify connections to both the mathematical and social justice goals of the lesson, you can also thoughtfully and positively impact student identity.

Student voice and choice in the opening and middle of the lessons are important components of ensuring equitable participation and learning. However, the closure or summary of each lesson is arguably the most important time because the teacher selects, sequences, and connects ideas from students. During this time, students begin to reflect on their personal or small-group ideas about the context and the mathematics, connect to others' ideas, and begin to solidify their understanding. As you help students to identify connections to both the mathematical and social justice goals of the lesson, you can also thoughtfully and positively impact student identity. When possible, you should

- Use the stories from students to highlight the mathematical and social justice question and exploration,

- Use the notes you kept while monitoring student exploration to purposefully select students to share their stories as they relate to the SJML and are meaningful to highlighting the mathematical discoveries, or

- Identify student ideas to be shared with the whole group in ways that promote the understanding of the main goals of the lesson.

Taking Action

An important element in the SJMLs in this book is the Taking Action piece. It is important for teachers to consider and make space for students to have opportunities to take action when making justice-oriented pedagogies a focus in their teaching. We asked that each lesson include a Taking Action component as a way for students to use the information that they were learning to express empathy, to recognize their own responsibility, to speak up with courage and respect, to make principled decisions about when to take a stand against bias, or to plan and carry out collective action. You may also want to consider different levels at which one could take action and articulate short- and long-term goals—at the individual, classroom, school, and community level, and beyond. Not every lesson will have multiple levels of action, but students should have opportunities to engage and to develop their sense of agency.

<aside>It is important for teachers to consider and make space for students to have opportunities to take action when making justice-oriented pedagogies a focus in their teaching.</aside>

Communicating With Stakeholders

As with any PreK–12 lesson, it is always wise to communicate regularly with important stakeholders, especially when teaching about social (in)justice. Because each context is unique, it is important for you to know the students in your classroom, their families, your administrators, and the greater community to anticipate how and what you should communicate about the lessons you teach.

In this section of the SJMLs, you will find suggestions of how a teacher might communicate about the social justice goals, the mathematics goals, and other pertinent information to families, administrators, and other stakeholders. Authors of the lessons might also include how students communicated their learning of both social justice content and mathematics to various stakeholders.

About the Author(s)

This section is provided so you will know a bit about the lesson author(s). Each author has provided a brief biography and describes how they became a social justice mathematics teacher.

CONCLUDING THOUGHTS BEFORE YOU GO TEACH

Upper elementary teachers have expertise in planning lessons that incorporate mathematics content and practices. You may also have experience with lessons that focus on social justice but do not necessarily include mathematics. A question raised earlier is relevant again here: How might you get started working at the

<aside>How might you get started working at the intersection of social justice and mathematical goals? We offer a recommendation: allow yourself to be a learner as you implement these SJMLs.</aside>

intersection of social justice and mathematical goals? We offer a recommendation: allow yourself to be a learner as you implement these SJMLs.

Focus first on the development of the social justice goal; be a problem-poser alongside your students. Also, resist the urge to force particular mathematical concepts or standards to fit with your social justice goal. Instead, view the landscape of mathematical knowledge as all potentially connected to the issue. The mathematics your students work on may not be aligned to where you are in the topics progression for the school year, but will connect to mathematics that they have learned or be a preview for mathematics they are yet to formalize. Second, as you listen to how students respond to the social justice learning goal, you likely will better understand the goal yourself.

We hope that these introductory chapters have formed a groundwork for you to further advance your efforts to TMSJ. We have attempted to provide frameworks and connect to strategies you already know well in order to create some comfort—both competence and confidence—to implement some of the lessons in this book, modified to fit the students in your classroom and the context in which you teach.

As we end this chapter, we send you forward to examine the SJMLs that come next in Chapters 5, 6, and 7. In this chapter, we described the SJML format we used in this book. Understanding this will be useful for your SJML implementation. However, more importantly, we hope that this lesson structure will propel you to develop high-quality mathematical investigations into issues of social injustice that the students in your classroom are passionate about and will engage them in investigation and action.

REFLECTION AND ACTION

It's time to dive in and teach a SJML. For some of us, it is a significant step—a sort of leap into trying out this kind of teaching and building on students' interests and connecting to families and communities. Others may already have experience doing this work, and these lessons may further enhance that work. The following steps can help you get started:

1. Return to the previous chapters, which ground the big ideas of TMSJ in upper elementary education. Commit to enhancing your teaching in alignment with these ideas, or frameworks, related to TMSJ. Teaching is, for us, the most important aspect of TMSJ.

2. Review the SJMLs in the next chapters and select one to teach in your class. Think through the lesson carefully, and modify as needed, recognizing that you might need to make changes to make the topic or questions a bit more relevant for your students, to make the mathematics more accessible or challenging, or to pose prompts to connect to local issues.

3. Invite a "critical friend," a trusted colleague, to help you think through, plan, and even watch you teach and help you reflect on the lesson.

As you begin or continue your journey to integrate SJMLs into your mathematics class, we invite you to share your story. Together, we can all make a difference in the lives of our students. A facebook group—Mathematics Lessons to Explore Injustice (https://www.facebook.com/groups/178098736933840)—offers a space to share your experiences implementing one of the lessons in this book. Please consider sharing your experiences with the lessons in this book or contributing anything else related to TMSJ. The lesson authors have shared thoughts—their successes, challenges, and advice—regarding this work, which we provide in Chapter 8.

PART
II

SOCIAL JUSTICE
MATHEMATICS LESSONS

CHAPTER 5

MATHEMATICS LESSONS FOR BUILDING AND EXAMINING IDENTITIES

In efforts toward equity and social justice, it is important to understand both one's own identity as well as the identities of others. The lessons that reflect this theme support students in learning about their own identities and privileges and in developing positive social identities. They also support students in learning about and valuing the identities of others and recognizing that multiple identities intersect to create unique and complex individuals.

LESSONS

Lesson No.	Lesson Title	Mathematics Focus Areas	Social Justice Topic	Authors
5.1	*Families Matter*	• Fraction Concepts and Operations	Family Structure Diversity	Nicky Meindl
5.2	*Playground Prejudice*	• Data Concepts and Statistical Thinking	LGBTQ+ Rights Bullying	Natalie Crist, Bryan Meyer, and John Staley
5.3	*Who Appears in Billboards?*	• Data Concepts and Statistical Thinking	Racial and Ethnic Disparities	Fernando Schlindwein Santino and Ana Carolina Faustino
5.4	*Family Story Problems*	• Fraction Concepts and Operations	Cultural Diversity	Sarah Ivey, Jami C. Friedrich, and Susan O. Cannon
5.5	*Exploring Mask-matics! Sociocultural and Environmental Concerns With Disposable Masks During COVID-19*	• Data Concepts and Statistical Thinking	Cultural Diversity Environmental Justice	Ho-Chieh Lin and Joanne Baltazar Vakil
5.6	*Challenging Ableist Assumptions in Mathematics Problems*	• Whole-Number Concepts and Operations	Disability Ableism	Courtney Koestler, Jennifer R. Newton, and Jan McGarry

LESSON 5.1 FAMILIES MATTER

Nicky Meindl

SOCIAL JUSTICE CONNECTION

In much of existing children's literature, family is portrayed as the nuclear family—made up of a cisgender husband, a cisgender wife, a cisgender daughter, and a cisgender son. This normative and dominant way of representing family structures is harmful and repressive to families that do not fit into this dominant perspective. This lesson explores the norms of family structures by asking students to examine family structures that they see both in their communities and the literature that they read. This lesson supports students' questioning about why these normative and dominant ways of representing family structures are often the only representations of families that students see in school curriculum. Students can take action toward diversifying the books in their school's library.

DEEP AND RICH MATHEMATICS

Students collaborate to gather data on the representations of families and family structures found in their set of school library books and create a data display to explain their results to the class. At the conclusion of the lesson, the entire class will share their findings to create a larger data display of everyone's data. The teacher and students create a line plot that represents the diversity, or lack of diversity, of family representations found in the books. As students explain their results, the teacher prompts them to explain their noticings, wonderings, feelings, and desire for action.

SOCIAL JUSTICE OUTCOMES

- I know my family and I do things the same as and different from other people and groups, and I know how to use what I learn from home, school, and other places that matter to me. (Identity 5)

- I like knowing people who are like me and different from me, and I treat each person with respect. (Diversity 6)

- I know when people are treated unfairly, and I can give examples of prejudice words, pictures, and rules. (Justice 12)

- I will work with my friends and family to make our school and community fair for everyone, and we will work hard and cooperate in order to achieve our goals. (Action 20)

MATHEMATICS CONCEPTS

- Understand a fraction as a whole partitioned into equal parts.

- Apply previous understandings of multiplication to multiply a fraction by a whole number.

- Solve word problems involving addition and subtraction of fractions.

- Draw graphs to represent data and solve problems involving addition and subtraction of fractions by using data presented in graphs.

Resources and Materials

- Book: *My Friends and Me* by Stephanie Stansbie

- Video: Ms. Katie reading *My Friends and Me* (https://bit.ly/3y4o6zd) or an alternative video of Ali Ayars reading this book (https://bit.ly/3Dv3Wj7)

- Worksheet 1: *Family Type Tally Sheet*

- Teacher Resource 1: *Sample Letter*

LESSON 1 FACILITATION

What Makes a Family?

Launch (20 minutes)

- Begin with a think–pair–share. Ask students to answer these questions with a quick-write or drawing:

  + *What makes a family?*

  + *What does a family look like?*

  + *Who is in your family?*

- Next, have students share their responses in small groups. Walk around the classroom during these discussions to note what common and unique ideas arise among the students.

TEACHER NOTE

You should consider the social identities of the students in your classroom, such as if students are adopted or in foster care, when discussing families. Many times, these students have no idea when, or even if, they will see their birth families, so it is important to remind them of what makes a family and what ways their foster/adopted guardians care for them like "blood family." Find and read books throughout the year that show the foster care system to give students mirrors to see themselves in as well as windows for others to learn about a piece of their experiences.

- Create a class concept map connecting the ideas that students have about families (see Figure 1 as an example).

Figure 1. Families Concept Map

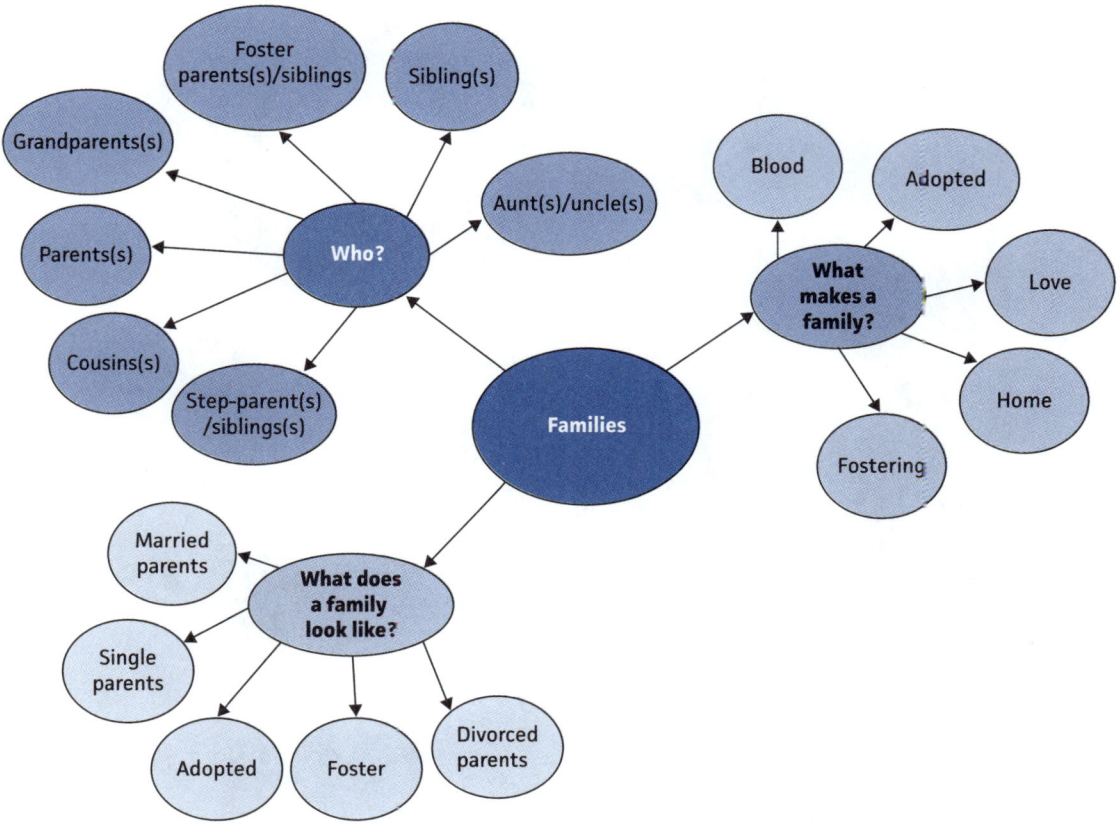

- Next, lead a discussion about the following questions:

  + *What families are usually represented in the books that you read?*

  + *What is your experience in reading books? Do you see your family represented in your reading?*

Explore (30 minutes)

- Read *My Friends and Me* by Stephanie Stansbie (or watch a video of a teacher reading the book on YouTube; see the *Resources and Materials* section) to the class. As you read, stop and draw attention to the different appearances of families in the text. Here are some questions to ask:

  + *What do you notice about this family?*

  + *Is this family the same as the previous family?*

  + *What is similar? What is different?*

  + *What cultures do we see represented here?*

- When you reach the end of the book, aid students in creating a chart to describe the text's different family structures (see Figure 2 as an example).

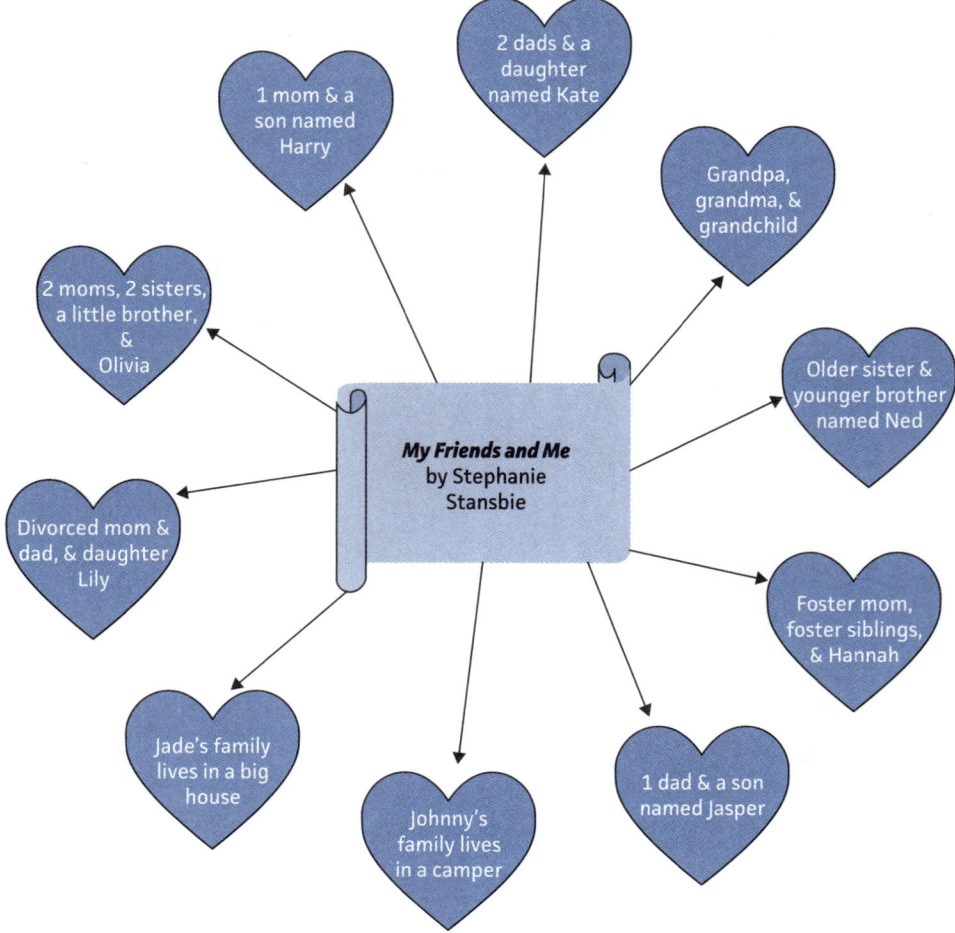

Summarize (15 minutes)

- Lead a whole-class discussion focused on these questions:

    + *What makes a family?*

    + *What are the different families we see represented in this story?*

    + *Who do we see in each of these families?*

    + *Do all families look the same? How do we know?*

    + *Does everyone in each family look identical?*

    + *Is every type of family represented in the book? How do we know?*

- Finally, draw students' attention to the wording on the second-to-last page, "Grown-ups . . . are fantastic at loving us." Ask students why they think the author chose to write "grown-ups" instead of "parents" here and how this is different from writing "parents are fantastic at loving us."

- Remind students to share a bit of new information that they learned in class with their families at home.

- As homework, have students find or create a picture of their family to share with classmates.

LESSON 2 FACILITATION

Examining Our Library Books

Before this activity, you will confer with your school or local librarian to determine the extent to which the librarian will assist you in the lesson. The focus will be on identifying which books are checked out most often by students from the library. Students will work in small groups to analyze the family structures represented in each book, so picture books will allow students to work more quickly.

Launch (10–15 minutes)

- Begin with this question: *How can we find out what types of families are represented in the books most read by students?* Allow students to share their ideas.

- Display the following goal on the board: "Gather and report on data of family structure representation in the most checked-out library books."

- Tell the class that today they will be looking closely at the pictures and mentions of families in this collection of books. Explain that when social scientists, such as sociologists, do this kind of work, they might go through the books a first time and develop their initial ideas about how to categorize the data; then they would meet and come up with a consistent approach they can use across all the different data sources, or books in this case. Tell the class that you will be doing the same thing. You will start by focusing on developing categories for types of families that everyone in the class can use. Then everyone will collect data for their books using those categories.

- Divide the students into small **heterogeneous groups** of three to four students.

- Each group will gather data from the library books about the family structures represented in some of the most popular (most checked-out) books in the library. If possible, the librarian can give a short presentation on the collection of most checked-out books.

- Tell the class that they will be looking closely at the pictures and mentions of families in the books for both primary and secondary characters. Tell the students that as they work, they need to come up with categories for the different types of families that they see in their books. Emphasize that they need the following for each category:

    + A definition or description (for example, "two moms" or "includes foster children")

    + An example from one of the books they are reviewing

Heterogeneous grouping of students—or grouping students with mixed abilities and strengths—is beneficial for students who historically have been underserved and marginalized in school as well as for those who consistently strive. It is beneficial for the learning of all students.

Explore (45–60 minutes)

- Have students work in groups to develop categories. Circulate and pose questions such as these:

    + *What categories did you create?*

    + *Who did you include in your categories?*

    + *Who is missing from your categories?*

- When you feel groups have a rich enough set of categories, bring them together for a whole-class discussion.

Summarize (20–25 minutes)

- Tell the class that each group will share their ideas and then you will work collectively to develop a common list of categories everyone can use. But first emphasize that this can be a challenging topic by saying:

    Analyzing diversity can be tricky for us and for social scientists because we want to celebrate all kinds of diversity in families. We don't want to get too focused on separating people or families into groups. But without analyzing the types of families we see, it can be hard to notice patterns. So we will come up with a way to analyze the data, but we also need to remind ourselves that the important part is valuing all kinds of people and families.

- Have each group briefly share their categories (along with examples from their books) and record them on the board. Encourage groups to question and comment on each other's categories, considering why they created these categories, what might be missing from their ideas, noticing unique ideas, and talking about similar categories that could be combined.

- Lead the class in combining similar categories and developing a common list for each group to use as they work. One key issue to discuss with the class is whether a family is allowed to be listed in two places. For example, one family may include both married parents and foster children. You will likely need to allow for overlaps, as developing exclusive categories for types of families will be too limiting. This could also be a place to tie into the use of Venn diagrams.

- Conclude by having students share pictures of their families with one another.

- Remind students to share a bit of new information that they learned in class with their families at home.

LESSON 3 FACILITATION

Representing Our Findings

Launch (10 minutes)

- Remind the class of the categories developed in the previous lesson. You can distribute Worksheet 1 (*Family Type Tally Sheet*) so that students can list family types and record tallies. As a scaffold, you may want to prepare tally sheets that list the family types in a table so students can record their tallies as shown here:

Family Type	Character Type	Tally Marks	Frequency
Family type 1	Primary		
	Secondary		
Family type 2	Primary		
	Secondary		
. . .			

- Tell students that today they have two jobs, first to collect their data and then to create a graph of their group's data. Remind them that there are many types of graphs, and they need to choose one that they think most clearly communicates their data. **Reminder:** *Do not tell students what kind of graph to use; let them decide what would work best for them.* You may want to briefly discuss pie charts and how they require non-overlapping (exclusive) categories. Depending on past work in your class, you may want to review different types of graphs. You can choose whether you will have students make their graphs by hand or if they will have access to Google Sheets or other graphing software. If students are working by hand, graph paper is a helpful tool.

Explore (30–45 minutes)

- Students work in their groups to create tally marks for each type of family that comes up in their books. If they encounter a family that they're not sure about, have them note where it is and save it so the rest of the class can discuss it after the data have been collected.

- When students are done, have groups discuss and then check in with you about how to represent their data with graphs. During this conversation, and later as you circulate and check in, pose questions such as these:

  + *Why did you choose this type of graph?*

  + *What are your labels?*

  + *What is your scale?*

  + *How does your graph tell a story about your data?*

Summarize (30–40 minutes)

- Have each group briefly present their graph to the class, answering the questions above.

- As a class, help the students combine their data sets and select one graph to make for the combined set of data. Discussion should focus on what graph they think will best represent the findings.

  Once the graph has been created, ask the students these questions:

  - *What does our class graph show us about the books in the library?*

  - *Does our class graph give us different information about the books in the library than each small group's graph? Why or why not?*

  - *Based on the information in our class graph, do we think that the most-often checked-out books in the library have a diverse representation of family structures? How do we know?*

  - *What types of families are well represented? What types are not well represented?*

  - *Why is there (or is there not) a diverse representation of family structures in the most checked-out books?*

  - *What can we do to have more diverse family structures represented in our library books? What can we do to encourage students to check out these books?*

- Remind students to share a bit of new information that they learned in class with their families at home.

LESSON 4 FACILITATION

Connecting Research to Action

This lesson provides an opportunity for students to connect their research to action. You can decide if you want to have different students or groups to take different forms of action or if you want to plan a common action (such as a presentation and/or letter-writing campaign) that everyone will participate in.

Launch (10–15 minutes)

- Ask the class to brainstorm ideas for how they can take action based on the research they have done so far. Here are some possible ideas:

  - Running a donation drive for books with diverse representations of families

  - Sharing results on social media and soliciting suggestions for book recommendations

  - Sharing results with family members

+ Sharing results with and/or writing letters advocating for change to the librarian, principal, school board, parent/caregiver teacher association (PCTA), or other members of the community

Explore (35–45 minutes)

- Students work on their selected form of action. Here we provide directions for a letter-writing campaign. The students will draft letters to their school's librarian, principal, PCTA, school board members, and/or district members. See Teacher Resource 1 for an example. The letters will advocate for more inclusive literature to be added to the school library by responding to the following questions:

    + *What is the importance of having diverse family structures represented in books?*

    + *What is the importance of adding more inclusive books to the library?*

    + *What are some books that could be added to the library? How do these books diversify the library through family structures?*

- The letters should be drafted, peer-edited, typed, and sent to their respective recipients.

Summarize (15–20 minutes)

- Groups of students share about how they connected research to action, why they chose that approach, and what they learned from the experience.

- Remind students to share a bit of new information that they learned in class with their families at home.

TAKING ACTION

As part of the end of the lesson, the students will write letters to the school librarian, school principal, school board members, or school district representatives to advocate for the addition of books that reflect the diverse family structures in both their community and their world. The students will detail why adding these books are important, provide suggestions for one or two books to be added to the school's library, and explain their reasoning for selecting the book.

As an extension activity, the students could work with their local and school libraries to highlight books that show diverse family structures. The students would be working with their librarians to create short summaries or book reviews to help entice other readers to select, read, and then share the book with others in their community. This could also include small art projects to create visuals to display about the highlighted books with diverse family structures, which would help draw attention to these books.

COMMUNICATING WITH STAKEHOLDERS

At the end of each day, the students will be prompted to share a bit of new information that they learned in class with their families at home. Additionally, the students' homework will be to bring in a picture or drawing of their family from home to share with the class in small groups, which honors the diverse family structures that are represented in the classroom community as well as helps students make connections between the similarities and differences among their families. Encourage families and parents/caregivers to visit the classroom to share their family's history and experiences. This connects to the Identity and Diversity domains of the Social Justice Standards, as students will be learning about their peers' families and comparing this to their own families.

ONLINE RESOURCES

 Available for download at **resources.corwin/TMSJ-UpperElementary**

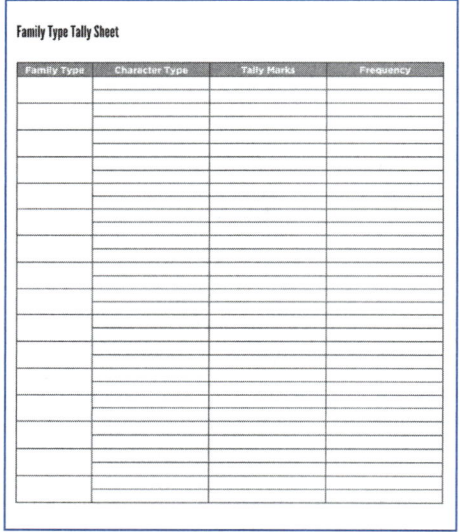

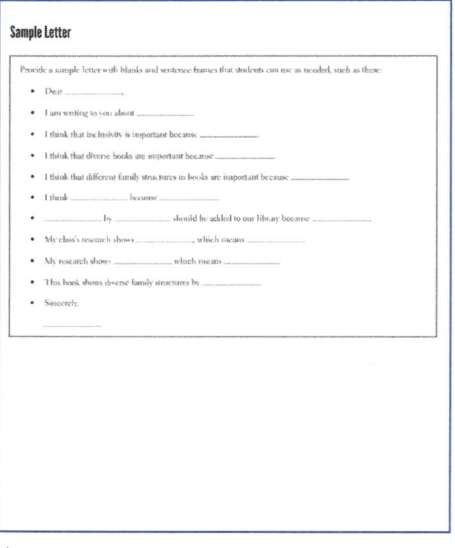

Worksheet 1: Family Type Tally Sheet

Teacher Resource 1: Sample Letter

ABOUT THE AUTHOR

Nicky Meindl is a queer, Chicanx, and nonbinary, first-generation graduate student in the Curriculum and Instruction Master's program at Chapman University while simultaneously student teaching with the purpose of becoming a social justice elementary educator. Their research focuses on the intersection of Queer Theory and Ethnic Studies in elementary schools.

LESSON 5.2 PLAYGROUND PREJUDICE

Natalie Crist, Bryan Meyer, and John W. Staley

SOCIAL JUSTICE CONNECTION

Name-calling and bullying are not new issues. They are part of a phenomenon that transcends all school levels. Bullying can lead to depression, loss of learning, low self-esteem, and, in extreme cases, suicide. It is important for students to understand that bullying and name-calling is a national issue that they and their classmates have experienced in their own lives. Helping students to identify the issue and understand the impact of name-calling is important to begin to impact the data and reduce the number of student suicides.

DEEP AND RICH MATHEMATICS

This lesson asks students to analyze and interpret data provided, collect data and create data displays, reason about the data and make generalizations and comparisons, and utilize numbers, operations, and fractions to operate on the data. The goal is for students to use the data to develop an understanding of the need to develop anti–name-calling materials, activities, and so on to make changes within their school community.

Resources and Materials

- Video: "Get Along Monsters: Don't Call Me Names" (https://bit.ly/3IsOWpG)

- Teacher Resource 1: *Survey Tool* (https://localsurvey.glsen .org/index.cfm)

- Worksheet 1: *Task 1—Notice and Wonder Data* (using data from the Gay, Lesbian and Straight Education Network (GLSEN) survey, "Playgrounds and Prejudice: Elementary School Climate in the United States," https:// bit.ly/3ItWMPF) (1 per student)

- Worksheet 2: *Task 2—Task Cards*

- Worksheet 3: *Task 3*

LESSON 1 FACILITATION
Experiencing Bullying

Launch (15 minutes)

- Begin by asking students, *Have you ever experienced a time when you were bullied by someone or heard someone being bullied?* Ask students to think about that memory and identify how it made them feel.

TEACHER NOTE

This may be a very sensitive subject for some students. As noted in Chapter 2, it is important that you have co-created a socially and emotionally supportive classroom environment. Also, focus on the behavior of bullying and not on an individual's identity. In this way, the focus is on the action of bullying—not labeling someone a bully—so as to support both the bullied student and to transform the behavior of the person identified in a story as bullying others (for more information, see Learning for Justice's "Bullying Basics" at https://bit.ly/32ZtPe9).

- Next show the video, "Get Along Monsters: Don't Call Me Names" (https://bit.ly/3IsOWpG).

- Hold a discussion about the actions and feelings in the video and in the students' lives. Identify the harm and the feelings. You may wish to record the students' ideas on a chart.

- Give your students a survey (see the *Survey Tool* in Teacher Resource 1, https://localsurvey.glsen.org/index.cfm) to respond to about bullying and why they are bullied, using the same questions from the GLSEN "Playgrounds and Prejudice" survey (https://bit.ly/3ItWMPF).

- Also, distribute to the grade level or school to gather data for Lessons 2 and 3.

  Note: You will need to compile these data into a spreadsheet or table for students to have for Lesson 2.

- Share with students that the class is going to look at how bullying occurs in other places beyond their school.

Explore (30 minutes)

- Begin by sharing the data from GLSEN "Playgrounds and Prejudice" (Worksheet 1, *Task 1, Notice and Wonder*). Give students a copy of the worksheet and display it for them to see as a class to hold the discussion.

- Using the Notice and Wonder Strategy and a T-chart, do the following:

  1. Give students about 5 minutes to observe the data and write down noticings and wonderings.

  2. Ask students: *What do notice and wonder about Graph A (pie graph)?* Record students' thoughts. If students are having difficulty, prompt them to look at the attributes of the graph such as color, categories, and numbers.

  3. Repeat the process for Graph B (bar graph). Prompt students if necessary.

  4. Ask students to think about their own experiences discussed in the launch: *How does your experience compare to the data presented in the chart? What conclusions could we make?*

  5. Have students create a list of questions that could be asked about the data. Or create questions that are (1) socially interesting and/or (2) lend themselves to appropriate arithmetic for the grade level. Then, the task becomes for students to calculate some values to answer those questions.

- If needed, you can use the following questions:

  + *How many students are in this school?*

  + *How many students report that bullying happens at their school?*

  + *What value do we get if we add up all of the numbers from the bar chart? What does that number mean? Why is it more than the number of students in the school?*

Summarize (15 minutes)

- Lead a discussion about the data. Begin by having students generalize based on the data. Consider the following questions:

  + *What conclusions can you make about how often students experience name-calling and bullying in school?*

  + *What would you predict we would find if we collected data in our school or district? Why do you think that?*

> + *What impact do you think that name-calling has on students in our class and school?*
>
> + *How do you think we could change the data and decrease the number of people who experience name-calling?*

LESSON 2 FACILITATION

Understanding and Representing Data

Launch (10 minutes)

- Display the graphs from Task 1 (Worksheet 1, *Task 1, Notice and Wonder*). Ask students to share what they learned from Task 1 and what generalizations they made.

- Tell students that they are going to learn about how data can be collected. Set the stage with an example. Say: *We have collected data in class before by doing a vote. Let's say we were going to choose what activity we want to have as our Friday reward, for instance. I am going to give you two choices. We are going to take the vote by closing our eyes and not looking.*

- Choose two activities that students like. Remind them: *You may only vote once.*

- Conduct the vote. Record the findings.

- Now say: *What if we were to do that again but this time we changed it a bit. What would happen if we didn't close our eyes when we voted? How would our vote change?* Ask students to think about how seeing what each other voted changes things or seeing what our friends vote for changes how we respond. Talk about how being anonymous changes the outcome and allows everyone to vote fairly.

Explore (60 minutes)

- Tell students that they are going to think about how to conduct a survey to collect data. Give groups the following task cards (also see Worksheet 2 for *Task 2, Task Cards*). This can be presented as a four corners activity, as a placemat activity, or for a group with assigned tasks. If facilitating as a placemat activity, put the cards on a large poster paper and assign a corner to each student in the group.

- Ask students to think about the following:

  + *How could we collect data from our school?*

  + *Would any of the following ideas work?*

  + *We are going to work in groups to discuss the proposed data collection methods.*

- Have students discuss the following two questions. They will need to come to group consensus and have a reason why they chose that method.

   1. *Each of the proposed survey methods has features that might lead you to get results that aren't accurate. What is wrong with each method?*

   2. *If your group had to choose one method to use for our class, which one do you think is best? That is, which one do you think will give us results that are most accurate?*

- Once groups have discussed, it is important to bring the class together to discuss the pros and cons of each of the methods.

TEACHER NOTE

In doing a survey, you are trying to get data that tell an accurate story about what people think. But there really is no such thing as a perfect survey. Sometimes, you can't actually ask every single person what they think. Other times, the method you use may lead people to not give honest answers.

- Share the data based on the survey questions provided in Lesson 1 (class, grade, school) in a pie graph and a table.

- Ask students to do a quick notice and wonder about the results.

- Give students the data collected in a spreadsheet or a frequency table with the item, the total population, and the response number as well as a variety of tools to create the graph or data display (see Worksheet 3, *Task* 3). Allow them to be creative to display the data. They can use physical materials such as hundred grids, Unifix cubes or unit cubes, graph paper, and so on to create visuals. They may also use more traditional approaches like a pictograph or bar graph.

- Give students the following instructions:

   Using the collected data (class, grade, school), create a data display to show the results. Before drawing, discuss the following questions with your group and be sure everyone agrees on answers.

      + *Do we need one graph or two?*

      + *What type of graph would be best to show this data?*

      + *What should the title for our graph be?*

      + *Will we have axes? If so, what will they represent? How should we scale them?*

- Continue:

When you feel ready, create your group's data display on the paper provided to you. Be sure that everyone in your group can explain your graph. You can practice by asking each other questions like these:

+ *How would you explain this graph to your teacher? To a family member?*

+ *How would you explain what each of the numbers means?*

+ *What are some questions that people might ask us about this graph?*

TEACHER NOTE

Students may display data as a traditional graph or they may use nontraditional displays to show quantities—like hundred grids or number lines—as a proportional amount.

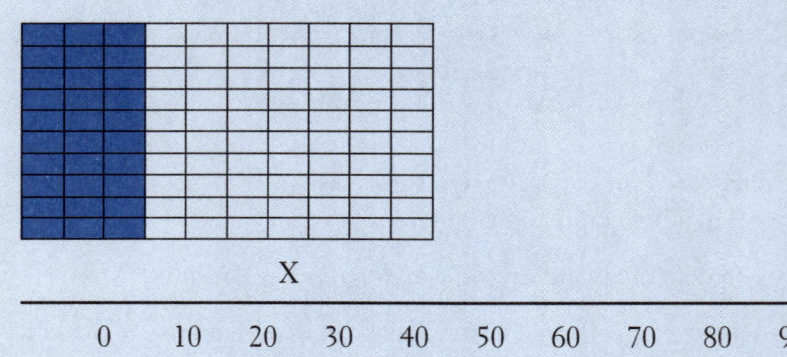

Summarize (15 minutes)
- End class with a gallery review using the graphs students created in Task 2. Have students leave sticky-note remarks about the graphs for each other. Save the graphs for Lesson 3 to open the discussion.

LESSON 3 FACILITATION

Taking Action

Launch (10 minutes)
- Share the following quote from the GLSEN "Playgrounds and Prejudice" elementary school climate survey:

Three out of every four elementary school students report that students at their school are called names, made fun of or bullied with at least some regularity (i.e., all the time, often, or sometimes).

- Ask: *When you read this quote, what comes to mind? How does that make you feel?* Have students share their thoughts.

- Use Unifix cubes of two colors to show the ratio 3 of one color and 1 of another color: 3 out of 4. Show how it can be written as a fraction as well. Students should see the amount written many ways, as it will help them generalize about their own data.

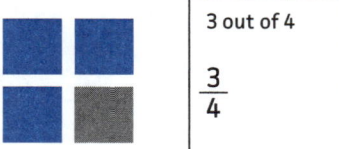

3 out of 4

$$\frac{3}{4}$$

Explore (60+ minutes)

- Lead a class discussion of the data displays. Given the data that you graphed yesterday, does your school have more or less bullying than the data presented in the GLSEN "Playgrounds and Prejudice" survey?

- Look at the graphs prepared by your class and discuss the following prompts:

  + *What data about your own school makes you sad or disappointed?*

  + *Why do you think people at your school might be doing this?*

  + *How does our data compare to the national data? What are similarities and differences?*

  + *What generalizations can we make about our data?*

  + *What are some things we could do to help stop this type of bullying?*

- You may wish to pick a piece of data and apply proportional reasoning as in the opening quote activity so students have a chance to think about the data in the same manner. This will extend the mathematics to increase the rigor of the activity.

- Give students time to take action against bullying by creating a statement, video, or poster. Additional activities may include creating a letter or a poster to share with the school. Let students be creative. You may wish to create a class contract or constitution that helps set parameters for students about being kind and support the antibullying message.

Summarize (15 minutes)
- Provide time for students to share their statements or show videos.

- Another option is to lead a discussion about what the students learned and how this will impact how they treat one another. This would be a great time to bring in the guidance counselor to talk about how to handle situations when students are being bullied. The teacher can also lead a discussion about how to handle seeking help when being bullied.

TAKING ACTION

You can give students time over multiple days to continue to develop an action. The goal is to do something with the new learning and have an impact on the data to decrease instances of bullying.

COMMUNICATING WITH STAKEHOLDERS

Students are encouraged in each lesson to consider what they can do to decrease bullying and how they might share their ideas with others and family members. It may be a way for students who are experiencing bullying or know of other students who are being bullied to gain the help and support they need. Lesson 3 focuses on the students taking action and sharing their suggestions with others by creating posters, making videos, or writing letters. Student-created items may be shared with school administrators, the student body, or community stakeholders.

ONLINE RESOURCES

 Available for download at **resources.corwin/TMSJ-UpperElementary**

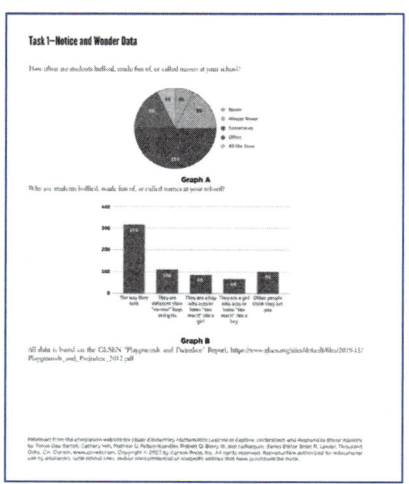

▲ Worksheet 1: Task 1—Notice and Wonder Data

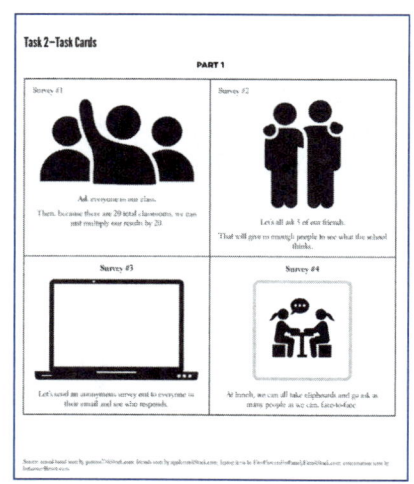

▲ Worksheet 2: Task 2—Task Cards

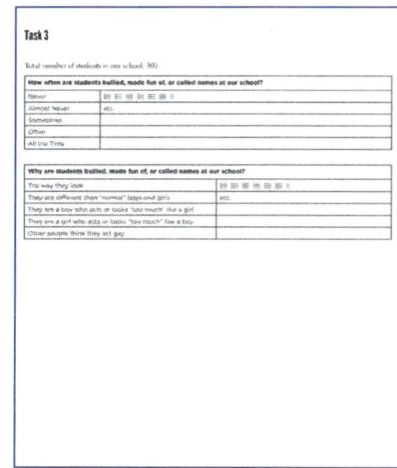

▲ Worksheet 3: Task 3

ABOUT THE AUTHORS

Natalie Crist has been an educator for 25 years and has served as a teacher, mathematics specialist, district mathematics leader, school principal, adjunct professor, and member of the NCSM board. Teacher education with an emphasis on culturally responsive instruction is at the forefront of her work.

Bryan Meyer has worked as a mathematics teacher, instructional coach, and mathematics program specialist. His current work focuses on school and district-wide systemic change efforts to change the structures and practices of mathematics education that protect the broader societal status quo through the marginalization of particular groups of students.

John W. Staley is currently the coordinator of special projects in Baltimore County Public Schools, where his primary work involves supporting schools in the continuous improvement process. He earned his bachelor of science in mathematics from the University of Maryland, College Park, a master's in secondary education from Temple University, and a PhD in mathematics education leadership from George Mason University.

- I know and like who I am and can talk about my family and myself and describe our various group identities. (Identity 1)

- I like knowing people who are like me and different from me, and I treat each person with respect. (Diversity 6)

- I know when people are treated unfairly, and I can give examples of prejudice words, pictures, and rules. (Justice 12)

- I will recognize my own responsibility to stand up to exclusion, prejudice, and injustice. (Action 17)

MATHEMATICS CONCEPTS

- Create data displays to organize, analyze, and communicate information.

MATHEMATICS PRACTICES

- Reason abstractly and quantitatively.

- Use appropriate tools strategically.

LESSON 5.3 WHO APPEARS IN BILLBOARDS?

Fernando Schlindwein Santino and
Ana Carolina Faustino

SOCIAL JUSTICE CONNECTION

The (in)visibility of marginalized groups based on race and ethnicity is the social justice topic in this lesson. The choice of ethnic groups that appear on billboards is not neutral, and thus can express prejudice. This lesson brings students' attention to representation in advertisements, enabling them to analyze and dialogue about the (in)visibility and selective visibility of certain racial and ethnic groups. Through data analysis, students can come to understand themselves as producers of mathematical knowledge and to recognize the influence of human beings in the production and analysis of mathematical data.

DEEP AND RICH MATHEMATICS

This lesson works with mathematical concepts related to data concepts and statistical thinking as students (1) collect, organize, represent, and interpret data, (2) calculate percentages, and (3) build bar charts, pie charts, and tables based on the photo registration of billboards in the city where students live.

Resources and Materials

- Worksheet 1: *Making Sense of Your Research*

- Photographic camera or cell phone (1 per student)

- Printed photos: total photos of billboards printed (1 set per group of 4 students)

- Paper (1 per student)

- Pen (1 per student)

LESSON 1 FACILITATION

Representation in the Media

Launch (30–45 minutes)

- The teacher begins by asking students the following questions:

    + *What is representation?*

    + *Why might it be important to be represented more in the media or, specifically, in billboard advertisements in our community?*

    + *What kind of advertisements do our local billboards focus on? Who are the people on the billboards? Do you remember their physical characteristics?*

Explore (15 minutes)

- After this dialogue, introduce the activity to students. Explain to students that they will be gathering a wide variety of advertisements in the local community to talk about representations. Give the following instruction (you may allow more time for students to take pictures if needed):

    + *You and your partners will have 1 week to take pictures of billboards in the city in which you live, considering the following criteria:*

        - It needs to contain people in the billboard images.

        - Each billboard must be photographed so that all of its contents are visible.

        - We may end up taking photos of the same billboard. To keep track of repetition in billboard images, the address or location of the photographed billboard must be noted. (An example of Figure 1 can be shared.) You can use Google Maps to record the location, and later you can also use the app when presenting the location of billboards to colleagues.

Figure 1. Location Record of the Photographed Billboard

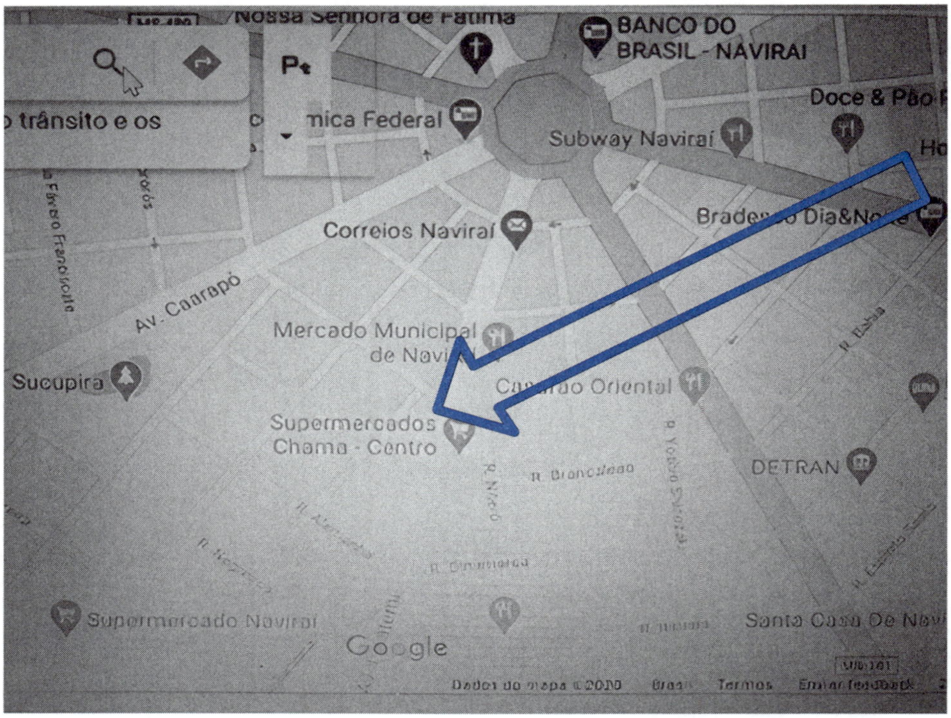

Source: Produced by the authors from Google Maps, 2019.

For example, students may find images like those in Figures 2 and 3 (from when this lesson was originally developed in Brazil).

Figure 2. Example of Photographed Billboard

Source: Researcher's file, 2019.

Figure 3. Second Example of Photographed Billboard

Source: Researcher's file, 2019.

Summarize (5–10 minutes)

- Discuss with students what they notice and wonder about the two examples presented. (Faces in billboards are blurred for publication purposes; students can and should see faces when sharing billboard images with one another.)

- Conclude the lesson by reminding students to collect pictures of billboards over the next week.

LESSON 2 FACILITATION

Interpreting Data

Launch (10–15 minutes)

- Say:

  I'm excited to see the pictures of billboards you have all collected. Today we are going to be analyzing these photos for patterns. First, let's reflect on the process. What was the experience like for you in taking photos? What sorts of things did you think about when selecting billboards?

- Have students share their thoughts and experiences.

Explore (multiday 45- to 55-minute sessions)

- Once students bring in the billboard images, this is the time to engage students in data interpretation. The goal at this point is to have students attend to who is visible in the billboards.

  + *Who is included in the billboard advertisements?*

  + *How are certain groups portrayed?*

  + *Does the billboard location matter?*

  + *Are there trends from one specific area to the next?*

- At this point, you can choose to have a class discussion and reach a consensus about the terms you will use for identifying people in the billboards (possibly informed by language used by governmental agencies, such as the U.S. Census) or you can allow groups to develop their own terminology.

- In any case, discuss the fact that the language used for race and ethnicity, gender, and other ways of identifying people is constantly shifting. Also raise the issue of the problem of making assumptions about people based on how they look. It is important to understand the data as how the students' *read* the people in the billboards. Depending on their prior experiences and the content of the billboards, students may come up with categories such as white, people of Color, Black, Asian or Asian American, American Indian, straight, gay, man, woman, transgender, parent (or mother or father), adult, child, (visibly) disabled, poor, rich, and so forth.

- Organize the class into groups of four students. Each group will analyze the total set of photos collected using tally marks to keep track of what they notice. Walk around to the groups to encourage students to consider the context of the advertising and the manner in which people are shown.

- During the process of categorization and identification of social identities in the billboards, ask the following questions:

  + *What are the similarities that you notice in the billboards? What are the differences?*

  + *Which racial and ethnic groups appear most frequently in the billboard images? Which racial and ethnic groups appear least often in the billboard images?*

  + *Let's use mathematics to identify the differences and similarities as well as to identify trends. What categories can be listed to help us to understand the data?*

  + *Now that the categories have been identified, create a table to record the categories and quantity (frequency). Remember to include a title, categories, quantity, and source.*

  + *Which group appears most frequently on the billboards? Why is that? What reasons can you think of?*

+ *Which group appears least on billboards? Why do you think this is so?*

+ *How do the findings make you feel? How do these images affect you personally?*

- Share with students Worksheet 1 (*Making Sense of Your Research*). Take time to discuss the terminology used by the U.S. Census (or another source you use), how it compares to the terms students developed, and overlaps among groups (for instance, in the U.S. Census Quick Facts data, "Hispanic or Latino" is considered an ethnicity and can overlap with the racial groups).

- The class can also create a whole-class graph or chart for better collaborative conversations. Ask: *What might be some advantages to a particular group of people if they are well represented in advertising? What would be some disadvantages?*

TEACHER NOTE

When we engaged in this activity with students in Brazil, the students calculated the percentage of the data, and the analysis was carried out over two class sessions. Subsequently, each of the groups shared their analysis. The data presented showed that 87.5% of the photographed billboards represented white people and only 12.5% represented Black people and no Indigenous people.

Students compared the results with the official demographic data of the Brazilian population in 2019, and they noticed a disproportion between the percentage of Black Brazilians of 56.2% and the percentage that refers to their visibility on the billboards, of 12.5%. Another highlight was the absence of Indigenous people on billboards. Another criterion explained by the students during the presentation of their findings was whether people present on the billboards were in a group or if they appeared alone. All the people who appeared alone on billboards were white. Black people always appeared on billboards accompanied by people considered white, which showed the invisibility of Black people, as they never appeared as central figures on billboards. The students showed that children are not born prejudiced or racist but learn in the environment in which they live.

Summarize (60–75 minutes)

- Representation might change significantly depending on the location of the advertisements. For example, groups well represented in an advertisement by a school might differ widely from groups well represented in the advertisements located in a bustling city center with restaurants and bars. In all forms of media, certain groups are represented more than others. Who is most seen? Who is least seen? Are there stereotypes or misrepresentations perpetuated? Have students determine what actions they would like to take based on their findings. Here are a few recommendations:

+ Have students create an infographic or video that summarizes their process of investigation, what the group observed on the billboards, their findings, and their learning. This text or video should be written in a way that helps you communicate their ideas to schoolmates and adults in the community.

+ Create your own billboard idea that would promote more equitable representation. Choose the characters, the type of advertising you want to run, and the possible location for this billboard, and think for whom this billboard is supposed to be communicating to.

TAKING ACTION

From the data analyzed, the students will create texts and videos (journal) to communicate their ideas to their community. Students will propose alternative billboards that will be designed by them. In addition, such findings can be communicated to the companies responsible for contracting the advertisements that are shown on the billboards as well as the companies that produce the billboards. This can be done by email, for instance.

COMMUNICATING WITH STAKEHOLDERS

Students will communicate their ideas through videos and texts (journal) about what they have found from the data collection and analysis. In addition, they will show other possibilities of billboards that promote more social justice. For that, the teacher can organize a sample of student videos, for instance. They can share these videos and texts (journal) with family members, community members, a local city council, and so on.

ONLINE RESOURCE

 Available for download at **resources.corwin/TMSJ-UpperElementary**

Making Sense of Your Research

Presenting results through graphs: Create charts with the data found by your group. You can create graphics with or without the use of technology. Think about the following questions to guide your creation:

- What kind of graph can you use to organize the information? List at least two types.

- What type of graph/chart would work best to display your findings? What type of graph/chart would support your claim?

- What kind of graph will you build? Why?

- What elements does a chart need to have to be understandable? Describe them.

- Choose a title for the chart and explain your reason for choosing this title.

- Build a legend regarding the representation of the data in the graph.

▲
Worksheet 1: Making Sense of Your Research

ABOUT THE AUTHORS

 **Fernando Schlindwein Santino** is studying for a master's degree in education from São Paulo State University in Brazil. He has a scholarship from the São Paulo Research Foundation. He has a bachelor's degree in pedagogy from the Federal University of Mato Grosso do Sul, Brazil.

 **Ana Carolina Faustino** is an associate professor in education at the Federal University of Mato Grosso do Sul in Brazil. She has a bachelor's degree in pedagogy and a master's degree in education, both from the Federal University of São Carlos in Brazil. She has a PhD in mathematics education from São Paulo State University in Brazil.

- I know about my family history and culture and about current and past contributions of people in my main identity groups. (Identity 2)

- I know my family and I do things the same as and different from other people and groups, and I know how to use what I learn from home, school and other places that matter to me. (Identity 5)

- I will respectfully express curiosity about the history and lived experiences of others and will exchange ideas and beliefs in an open-minded way. (Diversity 8)

- I try and get to know people as individuals because I know it is unfair to think all people in a shared identity group are the same. (Justice 11)

MATHEMATICS CONCEPTS

- Represent and solve story problems with fractions.

- Focus on real-world contexts for understanding fractions conceptually.

- Write mathematical story problems based on real-world contexts.

LESSON 5.4 FAMILY STORY PROBLEMS

Sarah Ivey, Jami C. Friedrich, and Susan O. Cannon

SOCIAL JUSTICE CONNECTION

This lesson engages students in sharing about their own family and cultural practices, meals, and interests (Identity 2 and 4; see Appendix D for a full list of domains and standards) and in learning about their classmates' lives and experiences (Diversity 8, Justice 11). These contexts are used as a foundation for introducing and developing fraction understanding. The lesson focuses on strategies that are beneficial for all learners, but that are of particular importance for emergent multilingual students in navigating the more language-intense emphasis in the Common Core State Standards.

DEEP AND RICH MATHEMATICS

This mathematics lesson shifts the focus from memorizing key words and looking for clues about operations to students thinking about the action in the problem. This strategy has helped students from a wide array of backgrounds, but especially emergent multilingual students. As a result of using this approach, students feel empowered to tackle word problems that they originally found difficult to solve, let alone read and comprehend. When students focus on the action, they are set up to use manipulatives or mathematical representations to directly model the problem (Carpenter et al., 2015).

Resources and Materials

- Book: *Children's Mathematics: Cognitively Guided Instruction* by T. Carpenter and colleagues (2015)

- Worksheet 1: *Sketch and Solve Template*

LESSON 1 FACILITATION

Sharing a Meal

Launch (10 minutes)

Start with a picture of people sharing a meal like the one here.

Source: rawpixel/iStock.com

- Pose the following to students:

    + *Imagine that you are going to eat a meal with family or friends.*

    + *Who would be there? Family members? Friends? Some of both?*

    + *What food would you eat? How would you share it?*

Explore (20 minutes)

- Give students 10 minutes to work independently to draw the people that they eat with and to list some of the foods that they eat and how food is shared among their family and friends.

- After students complete their drawings, have them discuss with their partner by taking turns asking each other:

  + *How is food shared in your home?*

  + *How is food shared at a friend's or another family member's home that you have been to?*

Summarize (10 minutes)

- Have some students share with the whole class. You may choose to have students share their partner's work or their own. As the class shares, ask them how their drawing is similar to and different from others who have shared.

- Be sure to highlight the following two points as students share:

  + Different types of food can be shared in different ways. For example, point out that mashed potatoes might be scooped out, and we can think of these in terms of volume, weight, or scoops; a platter of lasagna might be thought of as an area that we cut into smaller pieces; and apples are generally distributed as whole units, but they can be cut into smaller pieces.

  + There are different ways of deciding how to share food and what it means to share *fairly* among families and friends. In real life, people often do not share food in mathematically equal ways, so it is important to acknowledge the range of practices used and how this might differ from a mathematics problem where we sometimes assume an equal share.

LESSON 2 FACILITATION

Introducing Drawings as a Problem Solving Tool

Launch (10 minutes)

- Reflect on what students discovered the day before. Ask the class: *What did we draw yesterday? What did you discuss with your partner/group?*

- Tell the class: *Today we are going to work on using drawings to help us solve story problems about our lives. We are going to start with sharing food. Does anyone remember some of the different ways people share food in their home?*

- After students share a few ideas, explain the following: *Families and friends share food in many different ways. In the story problems we are going to work on next, we will assume that everyone always shares the food equally, so that everyone gets the same amount.*

- Tell the class:

  An important part of mathematics is representations, which are drawings to show important ideas in mathematics. Today we are going to work on using pictures to represent, or show, what is happening in a story problem and to help us find the answer. So, for your first strategy today, I want you to draw a picture that shows what is happening in this problem and shows how you solved the problem. If you finish, you can come up with a second strategy to solve the problem with or without a picture.

Explore (20 minutes)

- Pose a problem like one of the following examples, but with names and foods from your students' work on Lesson 1, and with the numbers adjusted to best fit your mathematical goals. You may want to prepare extension questions for students who finish early.

  + **Example 1 (Fair Share).** A family of 4 was sharing latkes that they prepared for Hanukkah. There were 10 latkes. If they want everyone to have the same amount and to use up all the food, how many latkes will each family member get?

  + **Example 2 (Comparison).** Eduardo's friends were sharing dinner. There were 8 people sharing a 9 × 13-inch pan of lasagna. The family and friends shared the lasagna equally so that each person got the same amount. Mahdia's family was also having dinner. There were 6 people at the meal sharing a platter of lasagna. The family shared the lasagna equally so that each person got the same amount. Both platters of lasagna were the same size. What do you notice about the two meals? Which student's dinner guests have larger servings of lasagna? How do you know?

- As students work, circulate and check in on their reasoning and drawings. Your goal is to select a few students who can share their work: discuss their strategy with them beforehand and tell them you are going to have them share their strategy when everyone is done. (See the following Summarize section for suggestions on how to select students.)

Summarize (10 minutes)

- Select students to share that highlight a range of ways of thinking relative to your mathematical goals. For instance, for the Fair Share example, the two strategies presented next show the answer as two equivalent fractions, and you can use these strategies to have a conversation about whether the people received the same amount of food in each situation:

  + Each person receives 2 whole latkes and 1 half latke.

  + Each person receives 10 fourths of a latke.

- As another example, if your students are new to drawing mathematical pictures, you may want to highlight examples that illustrate the problem differently:

    + Using rectangles versus circles

    + Explicitly include people in the drawing or not

    + Pictures with lots of detail versus minimalist diagrams

- You can use differences like these to pose questions to the class about mathematical representations:

    + *Does it matter if we change the shape we use? Why or why not? Is it okay if we change the shape because it makes it easier for us to solve the problem even if it doesn't match real life?*

    + *How much detail do we need when solving a math problem versus in art class?*

Optional Lesson Summarize: Introduce Sketch and Solve Template (10 minutes)

- If you feel your students would benefit from an additional structure, you can introduce the *Sketch and Solve Template* (Worksheet 1 in the *Resources and Materials*).

- Emphasize that this is an organizational tool that can help students keep track of their work. Select one of the strategies a student shared and demonstrate how they could put it in the template:

    + At the top, write the story problem they are solving.

    + Under **Sketch**, recreate the student's drawing.

    + Under **Important Numbers/Information**, you list key information from the problem. For instance, using Example 1, you might write "4 people sharing 10 latkes equally."

    + Under **Solve**, students write their answer and explain how they got it using their sketch. With Example 1, they might write that each person received 2 whole latkes and 1 half latke. Encourage students to use highlighters or arrows to show where this solution comes from in the sketch.

Additional Practice: Summarize (10 minutes)

- You can repeat Lesson 2 as often as fits your instructional goals before and/or after continuing to Lesson 3. If you have introduced the *Sketch and Solve Template*, then you should make this available to students when they work.

- In addition to the discussion prompts in Lesson 2, it can be valuable to discuss the importance of connecting mathematics to students' lives. You can revisit this conversation a few times throughout the year with prompts like these:

  + *How does it make you feel when we do math problems about you or people in your class?*

  + *How do these problems make you feel compared to math problems from a book?*

  + *Are there other topics you would like to solve math problems about?*

  + *What have you learned about your classmates through these problems?*

LESSON 3 FACILITATION

Students Write Problems

You may wish to repeat this lesson at different points in the year (you do not need to repeat the Initial Launch each time if you have sufficient ideas to work with). As with Lesson 1, take notes on what students share so that you can refer to them later as you create problems in the future. The goal of this lesson is to broaden the conversation about students' lives to include other areas. This can be supplemented with one-on-one or small-group conversations and/or surveys with your students and/or their family members.

Initial Launch (20–30 minutes)

- Remind the class about your conversations from Lessons 1 and 2 where you discussed how different families and friends come together to share meals. Tell the class that today you want to brainstorm about other things that are important to them outside of school. Choose some or all of the following to list on the board (and feel free to add your own ideas), and explain that you are interested in learning about them as people and in learning about their families and friends:

  + Hobbies, sports, and activities

  + Traditions and holidays

  + Things people collect

  + Skills or jobs

- As students share ideas, ask follow-up questions and encourage other students to ask follow-up questions and make connections. As the teacher, one of your goals is to understand how to mathematize the examples students are providing. For example, if a student shares a board game their family likes to play, you might ask about the number of pieces or the scoring, so you are able to make realistic story problems about the context.

- Moving forward, as you repeat Lesson 2, you can pose a variety of story problems that touch on different areas of mathematics and connect to different parts of your students' lives. Your goal should be to include all students' experiences or interests throughout the year.

FIELD TESTER REFLECTION

A lot of discussion of hobbies. Students loved sharing . . . I see this being a great way to launch problem solving for kids in the fall. This is a great way for students to learn about each other and see there are a variety of cultures and traditions students in class may have. It is also a great way for the teacher to learn about students' interests.

—Heather Askay

Main Launch (20 minutes)

- Remind the class of the past conversations you have had in Lesson 1 and the Initial Launch of this lesson where you learned about the students' meals and families. Ask for a few volunteers to remind the class of something they learned about a peer or something about their family. Depending on how long it has been, you may want to share some examples from your notes from those days.

- Tell the class:

  A lot of times in math we solve problems that other people give us, but an important part of mathematics is posing, or creating, your own math problems. Today you will work in pairs to write your own story problems. When you write a story problem, I want you to think about what information you need to include and what information the problem solver must figure out or decide for themselves. Use examples of things that are important to you or your family.

- Many students will not be used to writing story problems. Provide the students with an example of a type of problem they have worked on recently. One fairly accessible problem type is a Fair Share problem: *A family of 4 was sharing latkes that they prepared for Hanukkah. There were 10 latkes. If they want everyone to have the same amount and to use up all the food, how many latkes will each family member get?*

- Ask the class what information is given in the problem (there are 4 people sharing 10 latkes) and what they are trying to find out (how much food each person gets). Then repeat this process with a second problem of the same type, but with different numbers and a different context. Ask what's the same and different about the two problems. In particular, point out that (1) both problems tell you the number of people and the amount of food (or stuff) they are trying to share, and (2) they both ask you to figure out how much each person gets. You can choose a different problem type if it fits better for your classroom.

Explore (40–60 minutes)

- Students work in pairs to write their own story problems and to solve them using a picture.

- Meet with students to give feedback, checking to be sure the problems are understandable, solvable, and related to the lives of the students or class.

- If time allows, have pairs swap partners and solve their problems. Alternatively, you may select problems to use for future Additional Practice lessons.

Summarize (10 minutes)

- Depending on the progress the class makes, the summary portion may come on a future day.

- Ask the class questions about creating their own problems and about solving other pairs' problems:

  + *What kinds of things make story problems easier to understand? What made them harder to understand?*

  + *What made the problems easier or harder to solve?*

  + *What did you think about when you were writing your problems?*

TAKING ACTION

Students might explore the idea of sharing food in different contexts, such as helping in the school cafeteria or soup kitchen sharing out the food. Students might also explore how food or other resources get shared in their community. They can consider what they learn about these practices in relation to their own ideas about sharing and fairness, and they can dialogue with community partners about their thinking.

COMMUNICATING WITH STAKEHOLDERS

Ask for a few volunteers to remind the class of something they learned about a peer or something about their family in this lesson. You may want to share some examples from your notes on those days. In addition, post the different story problems that students created as part of Lesson 3.

- Students select a problem to solve. They solve their problems and talk to the student(s) who created it to learn about the context. They should ask the authors questions like these:

  + *Why did you write your problem about _____?*

(Continued)

(Continued)

> + *Is _____ important to you or your family? Why?*
>
> + *What else do you want me to know about _____?*
>
> • Students take notes so that they can share both their mathematical thinking and what they learned about their classmates with a family member for homework.
>
> • Students share their selected problem with a family member for homework.

ONLINE RESOURCE

 Available for download at **resources.corwin/TMSJ-UpperElementary**

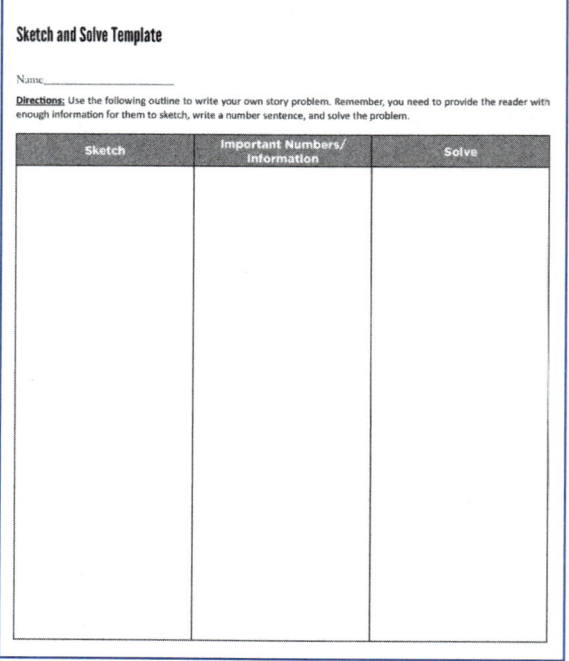

▲
Worksheet 1: Sketch and Solve Template

ABOUT THE AUTHORS

Sarah Ivey is a third-grade teacher at Oak Grove Elementary in Peachtree City, Georgia. Mathematics is her favorite subject to teach, and she is a proponent of math talk and increased collaboration in mathematics for all students. It has become Sarah's personal mission to make each one of the students in her class believe they can do math, even if they feel like they can't do it at first. She regularly works with emergent multilingual students, students with IEPs, and students who have had few opportunities to learn and engage with mathematics. After just the first few weeks of each school year, you will hear her students referring to themselves as mathematicians. It is one of the most beautiful and rewarding experiences Sarah has ever been a part of.

Jami C. Friedrich is a postdoctoral fellow with the Hines Family Foundation in Atlanta, Georgia. She has been a middle and high school mathematics teacher and coach. Her research interests involve culturally responsive pedagogy and STEM education with underserved and underrepresented populations.

Susan O. Cannon is an assistant professor of early childhood/middle grades education at Mercer University. Susan taught in public elementary and middle schools in Atlanta, Georgia, for 13 years before pursuing her doctoral degree. She loves learning with and from preservice and in-service teachers.

SOCIOCULTURAL AND ENVIRONMENTAL CONCERNS WITH DISPOSABLE MASKS DURING COVID-19

Ho-Chieh Lin and Joanne Baltazar Vakil

SOCIAL JUSTICE CONNECTION

Since early 2020, the COVID-19 virus has caused widespread panic and disruption around the world. To combat the pandemic, the government and health authorities of many countries and U.S. states mandated that their citizens wear face coverings indoors and if not vaccinated, as they were proven to effectively control the spread of the virus. This lesson provides a space for students to have courageous conversations about masks, including why we wear them, how we all might not be able to afford them, and how we each make an impact on our environment when we dispose of them.

DEEP AND RICH MATHEMATICS

Students begin this lesson by exploring concepts of data analysis and examining variability in mask-wearing by regions and demographics. Data analysis continues and includes percent concepts and operations as students examine the environmental impact of single-use disposable masks and consider more sustainable solutions.

Resources and Materials

- Article and Graph: "More Americans say they are regularly wearing masks in stores and other businesses," by Stefanie Kramer, *Pew Center Research*, August 27, 2020 (https://pewrsr.ch/31ua2U3)

- Article and Graph: "Wear a mask? Yes, always wear a mask," *Institute for Health Metrics and Evaluation*, June 18, 2020 (https://bit.ly/339K8Ft)

- Video: "Pandemic Pollution: Disposable Masks, Gloves Are Saving Lives But Ruining the Environment," by Stephanie Sy and Lorna Baldwin, *PBS News*, April 22, 2021 (https://to.pbs.org/3IwbPs7)

- Article: "Where did 5,500 tonnes of plastic waste from masks end up?" by Jenny Yeh, *Greenpeace*, August 14, 2020 (https://bit.ly/3rJLelx)

- Article: "Three million masks every minute: how Covid-19 is choking the planet," *The Straits Times*, January 9, 2021 (https://bit.ly/3ow2pF2)

- Website: Centers for Disease Control and Prevention (CDC) stance on face coverings (https://bit.ly/339xshV)

- Worksheet 1: *Mask Disposal Data Collection Sheet*

Optional Resources

- Article: "Living with facemasks: How to stow them, reuse disposables and more," by Emily Chung, *CBC News*, August 6, 2020 (https://bit.ly/3rIJzg0)

- Article: "All about the Coronavirus," *Junior Scholastic*, September 18, 2020 (https://bit.ly/3IvytB6)

- Article: "Face masks: What the data say," by Lynne Peeples, *Nature*, October 6, 2020 (https://go.nature.com/3dtq2b7)

- Article: "Face masks: New solutions to reduce their negative impact on the environment," by Abigail Saltmarsh, *Medical Expo*, January 6, 2021 (https://bit.ly/3EunCVw)

- Article: "How to stop discarded face masks from polluting the planet," by Laura Parker, *National Geographic*, April 14, 2021 (https://on.natgeo.com/3dAhGyh)

- Family Resource: "Coronavirus (COVID-19): Kids and Masks," Nemours Children's Health (https://bit.ly/31Ewf1p)

- Poster: "All About Masks," *Sesame Street* (https://bit.ly/3y6XwWb)

- Poster: "Be a Mask Hero," *Junior Scholastic* (https://bit.ly/3GsVa7f)

- Tutorial: "How to Make a Facemask," *CBC Kids News* (https://bit.ly/3drK85J)

- Video: "Covid-19: Does Your Kid Really Need a Mask?" *NOVA* (https://to.pbs.org/3y7ywOO)

- Website: Mayo Clinic Network, "Benefits of Kids Wearing Masks in School." (https://mayocl.in/3rLrxKg)

MATHEMATICS PRACTICES
- Make sense of problems and persevere in solving them.
- Reason abstractly and quantitatively.
- Construct viable arguments and critique the reasoning of others.
- Use appropriate tools strategically.

LESSON 1 FACILITATION

Identities Around Masks

Launch (20 minutes)
The History and Culture of Masks

- Begin the discussion of mask-wearing during the COVID -19 pandemic with two graphs. The first graph is from the Pew Center Research article (August 7, 2020), "More Americans say they are regularly wearing masks in stores and other businesses" (https://pewrsr.ch/31ua2U3). As the COVID-19 pandemic continued, the Pew Research Center conducted surveys from June 4 to 10 and August 3 to 16, 2020, asking Americans if they were regularly wearing a mask or face covering in stores and other businesses. The second graph is from the Institute for Health Metrics and Evaluation website, which is a population health research center at the University of Washington (https://bit.ly/339K8Ft). This graph uses a color map to depict the percentage of U.S. people surveyed on June 13, 2020, who said they always wear a mask when going out.

- With the graph (see Figure 1), ask students to share their thoughts on these questions:

  + *What do you notice?*

  + *What do you wonder? What trends do you see?*

  + *Do you think this has continued?*

  + *What possible reasons do you think have caused the differences among groups or regions?*

- Often children will share the following noticings:

  + *More Americans were wearing their masks in August 2020 than in June 2020.*

  + *More than half of Americans were wearing masks in June 2020.*

  + *The information does not include people younger than 18. Why are we not included?*

  + *More people wore masks in some U.S. states than in other states on June 13, 2020, at both time points. Why is that?*

Figure 1. Graphs of Mask Usage in 2020

Mask use increased in summer months

% who say that, in the past month, they've worn a mask or face covering when in stores or other businesses all or most of the time

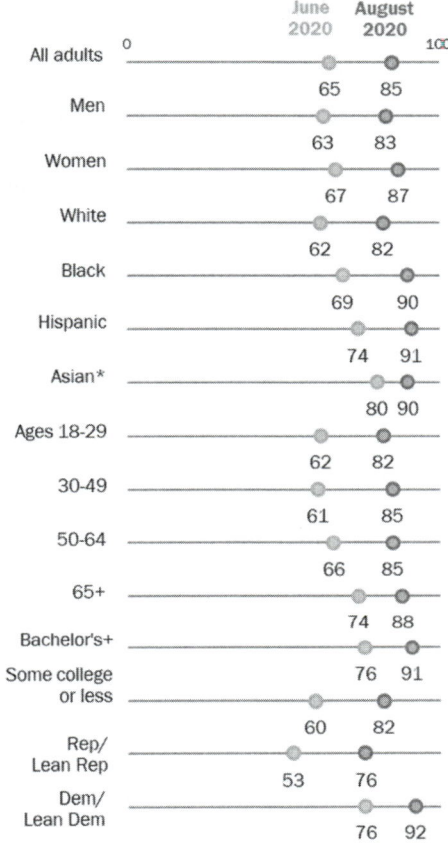

	June 2020	August 2020
All adults		
Men	65	85
Women	63	83
White	67	87
Black	62	82
Hispanic	69	90
Asian*	74	91
Ages 18-29	80	90
30-49	62	82
50-64	61	85
65+	66	85
Bachelor's+	74	88
Some college or less	76	91
Rep/ Lean Rep	60	82
Dem/ Lean Dem	53	76
	76	92

*Asian Americans were interviewed in English only.
Note: White, Black and Asian adults include those who report being only one race and are not Hispanic. Hispanics are of any race.
"Some college" includes those with an associate degree and those who attended college but did not obtain a degree.
Source: Surveys of U.S. adults conducted June 4-10 and August 3-16, 2020.

PEW RESEARCH CENTER

Percent of people who say they always wear a mask when going out, June 13

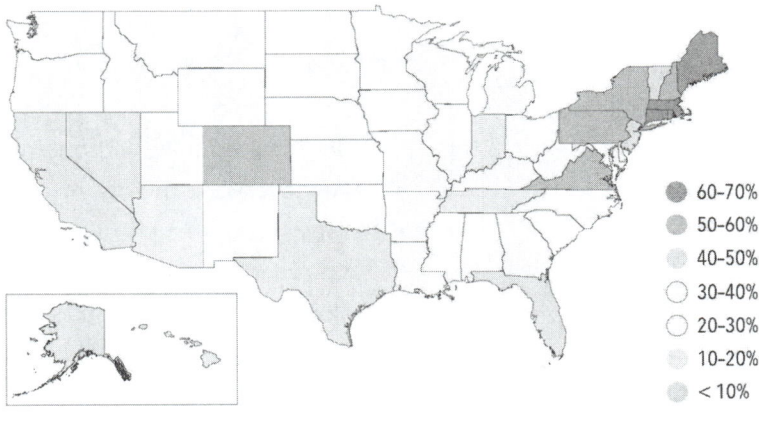

- 60–70%
- 50–60%
- 40–50%
- 30–40%
- 20–30%
- 10–20%
- < 10%

Data source: Premise

Source: Institute for Health Metrics Evaluation. Used with permission. All rights reserved.

TEACHER NOTE

The word *percent* comes from the phrase *per cent*. Cent is a root that means one hundred, so per cent literally means per one hundred. Percents may be a new concept for students. Use 10 × 10 grid paper and base-10 blocks (using the 1 flat to represent 100%, 1 rod to represent 10%, and 1 unit to represent 1%) to serve as visual and physical resources to support sensemaking of the quantity.

- The pandemic has impacted everyone; however, students may have differing experiences with and background knowledge about the pandemic, mask-wearing, and other forms of prevention. Give students the opportunity to discuss different opinions about why some people wear masks and some don't. When differing opinions occur in the classroom, write down student ideas and allow time for further research, supporting students to consider the credibility of sources. Remind students that part of being a mathematician involves constructing viable arguments and critiquing the reasoning of others (Standard for Mathematical Practice 3). However, this should be done in a civil, respectful manner. Review class norms established during the year, such as respecting others, disagreeing with ideas not people, listening to others' ideas, and so forth. See the sample class discussion guidelines in Figure 2. The following websites can also serve as references:

  + Poster: "All About Masks," *Sesame Street* (https://bit.ly/3y6XwWb)

  + Poster: "Be a Mask Hero," *Junior Scholastic* (https://bit.ly/3GsVa7f)

  + Article: "Face masks: What the data say," by Lynne Peeples, *Nature*, October 6, 2020 (https://go.nature.com/3dtq2b7)

  + Article: "All about the Coronavirus," *Junior Scholastic*, September 18, 2020 (https://bit.ly/3IvytB6)

Figure 2. Sample Class Discussion Guidelines

- Listen respectfully, without interrupting.
- Allow everyone the opportunity to speak.
- Always use a respectful tone.
- Criticize ideas, not individuals or groups.
- Ask questions when you don't understand; don't assume you know others' thinking or motivations.
- Try to see the issue from the other person's perspective before stating your opinion.

Explore (40 minutes)

Masks can protect human health and human lives, but they can do harm to animals and our environment if we are not conscious of their environmental toll. Show the following pictures of a trapped animal from a news website:

Source: Neurone89/iStock.com

- Then ask the following questions:

  + *What do you notice in the picture?*

  + *What is the negative environmental impact of discarded masks? Why?*

- Possible student responses:

  + *An animal is trapped by mask strings . . .*

  + *Masks that are thrown away can end up hurting animals . . .*

  + *The masks keep piling up . . .*

  + *I don't think masks are biodegradable, or if they are, they don't degrade fast enough . . .*

  + *The masks hurt/endanger the animals.*

- As a class, watch the video, "Pandemic Pollution: Disposable Masks, Gloves Are Saving Lives But Ruining the Environment" (https://to.pbs.org/3IwbPs7).

- Then ask the following questions:

  + *What did you learn from the video?*

- Possible student responses:

  + *Some of the animals eat masks and that is bad for them . . .*

  + *The animals get tangled in the straps of the masks and that is dangerous.*

- Summarize the video's lesson by pointing out major risks of mask waste. Resources can be drawn from the Greenpeace article, "Where did 5,500 tonnes of plastic waste from masks end up?" (https://bit.ly/3rJLelx).

- Start the next mathematics activity by asking students to estimate the impact of mask waste with mathematics. Post the following prompt on the whiteboard for students:

  Every minute of the day, we throw away 3 million face masks. How many masks are thrown away in half a minute? In 15 seconds? In 10 seconds? In a second? In 10 minutes? In an hour? In a day?

TEACHER NOTE

Three million is a very big number and can feel abstract to students. If needed, support students to develop a sense of scale of millions with smaller quantities that are more familiar. The progression of questions as follows can serve as a number string to support students' problem solving through mathematical reasoning:

- 1 minute: 3 million face masks discarded

- $\frac{1}{2}$ minute (30 seconds): 1.5 million or 1,500,000 face masks discarded

- $\frac{1}{4}$ minute (15 seconds): 750,000 face masks discarded

- $\frac{1}{6}$ minute (10 seconds): 500,000 face masks discarded

Provide mathematics resources (e.g., place value charts, arrow cards, base-10 blocks) to support students' understanding of place value concepts.

The following questions can be asked to support student problem solving and reasoning:

- *How can knowing the number of masks discarded in $\frac{1}{2}$ minute (30 seconds) help you figure out the amount of masks discarded in $\frac{1}{4}$ minute (15 seconds)?*

- *How might you use what you figured out earlier (number of masks discarded in $\frac{1}{2}$ minute or $\frac{1}{4}$ minute) to figure out $\frac{1}{6}$ of a minute?*

Periods	Billions			Millions			Thousands			Ones		
Places	Hundred Billion	Ten Billion	Billion	Hundred Million	Ten Million	Million	Hundred Thousand	Ten Thousand	Thousand	Hundreds	Tens	Ones
Numbers												

Summarize (10 minutes)

- End class with students sharing their problem-solving strategies for the earlier problems. Afterward, ask students to write on a sticky note to respond to one of the following prompts, which will support the next day's activity:

    + *What was one fact that stood out from today's investigation?*

    + *What surprised you about today's investigation with mask waste?*

    + *How do we contribute to mask waste and what are some possible ways to reduce its negative impact on the environment?*

LESSON 2 FACILITATION

Environmental Justice Issues Concerning Mask Disposal

Launch: Act I (20-minute introduction, followed by 2 weeks of at-home data collection)

Household Mask Use

- Share the following facts from *The Straits Times* article, "Three million masks every minute: How Covid-19 is choking the planet" (https://bit.ly/3ow2pF2): *Experts now estimate that each month, 129 billion face masks and 65 billion gloves are used and disposed of globally.*

- Ask students to think about the following prompt: *How much do we contribute to the use and disposal of masks individually and as a class?*

- Give students Worksheet 1 (*Mask Disposal Data Collection Sheet*) and instruct them to record their household use of disposable masks for 2 weeks, including the time they throw out old masks.

Explore: Act II (60 minutes)

Class Analysis

- Based on each student's facemask use log, have them create a line graph to show changes in the household mask use over a span of 2 weeks. The *x*-axis has the days of the weeks to capture the time period, and the *y*-axis has the number of masks discarded daily.

- Have students work in pairs and share their line graph documenting their household use of masks in the past 2 weeks.

- Ask student pairs to discuss the following questions to begin considering mask consumption and ways toward waste reduction.

    + *How many disposable masks did your family discard in a week?*

    + *What trends do you notice?*

    + *Are there differences between you and your partner?*

+ *Why were there differences?*

+ *What were some of the ways that led to decreased mask waste?*

Summarize: Act III (60 minutes)
Taking Actions on Mask Waste Reduction

- Use the following question to facilitate whole-class discussions building from the earlier partner activity to support student action on mask reduction:

What can people in our class and others do to reduce the waste from disposable masks and its impact on our environment?

Below are possible student responses with follow-up teacher questions to support investigation toward possible actions on mask waste reduction.

Cloth Mask Solution

- **Student response:** *I noticed my partner uses multilayered cloth masks instead of disposable masks.*

- **Possible teacher prompts:** Ask students to consider the environmental burden in cloth masks in comparison to disposable masks.

  + *What are environmental burdens to cloth masks?*

  + *How much water is needed to clean the masks daily?*

  + *What are other ways to disinfect masks?*

Sustainable Mask Solution

- **Student response:** *I want to research more about sustainable masks.*

- **Possible teacher prompts:** Ask students to give examples of sustainable objects.

  + *What makes something sustainable?*

  + *What does it mean to be eco-friendly? How can cloth masks and disposable masks be sustainable?*

  + *What are the benefits of sustainable masks?*

Proper Disposal of Masks

- **Student response:** *I will pick up masks on the street with gloves and put them in the trash can or I will cut off mask strings when discarding a mask.*

- **Possible teacher prompts:** Complement student's consideration and ask for details.

  + *I like how you're considering proper disposal of masks. What are things we need to consider to safely and properly dispose of masks?*

Reusing Disposable Masks

- **Student response:** *I want to reuse my disposable masks instead of throwing them away each day.*

- **Possible teacher prompts:** Help students consider impacts of their actions.

    + *Are there safe ways to reuse disposable masks?*

    + *What may be the impact of reusing disposable masks on our own safety and the environment?*

- Students may work in triads and quads to explore mask waste reduction strategies and have them create a poster with their findings. The poster must include answers to the following:

    + *How does your action reduce the environmental impact of disposable masks?*

    + *What data supports your recommended action? Explain the data in words and pictures so that it can be understandable for all K–8 students at the school site.*

Here are some additional resources you may use to support student investigation of mask waste reduction:

- Provide the article "Living with facemasks: How to stow them, reuse disposables and more," by Emily Chung, *CBC News*, August 6, 2020 (https://bit.ly/3rIJzg0). According to experts, reusing surgical masks that have not been used in high-risk areas is a possible solution. This is because viruses can be destroyed under sunlight or long exposure to the natural environment. Therefore, one suggested action plan is to take turns using five to seven masks. For example, if you use a mask on Monday, then leave it in a clean paper bag for 3 to 4 days before you use it next time.

- Additional recommendations are provided in the article "How to stop discarded face masks from polluting the planet," by Laura Parker, *National Geographic*, April 14, 2021 (https://on.natgeo.com/3dAhGyh)

- See the article "Face masks: New solutions to reduce their negative impact on the environment," by Abigail Saltmarsh, *Medical Expo*, January 6, 2021 (https://bit.ly/3EunCVw), for recommendations to reduce the negative impact on the environment from zero waste, creating protective gear, to biodegradable mask options.

TAKING ACTION

Give students time to take action for mask waste reduction by sharing their learning with others through creating a statement, video, or poster to share with the school. Let students be creative. Give them time over multiple days to develop a

(Continued)

(Continued)

community action. The goal is to do something with the new learning and have an impact on the data to decrease mask waste. Here are some suggestions:

- Design and make a reusable mask following the *CBC News* "How to Make a Facemask" tutorial (https://bit.ly/3drK85J).

- Create an infographic about information learned during the school's weekly notice or principal announcement.

- Partner with local community organizations and share their resources with families and the school community.

COMMUNICATING WITH STAKEHOLDERS

Mask-wearing can be a controversial topic in some schools. Encourage local medical providers to visit the classroom to share the benefits of mask-wearing. Provide family-friendly and age-appropriate resources to share with parents/caregivers. Below are a few:

- "Benefits of Kids Wearing Masks in School" by Deb Balzer (https://mayocl.in/3rLrxKg): This article from the Mayo Clinic Network addresses any negative and positive effects of children wearing masks in school.

- "Coronavirus (COVID-19): Kids and Masks" (https://bit.ly/31Ewf1p): This resource from Nemours Children's Health can be sent to parents before the lesson is taught as a means to inform and educate families about topics on why people should wear masks, how masks help, and who should wear masks.

- "Covid-19: Does Your Kid Really Need a Mask?" (https://to.pbs.org/3y7ywOO). This *NOVA* video debunks some mask myths and likens masks to a protective measure such as helmets. Habits such as physical distancing and avoiding touching face coverings are also addressed.

ONLINE RESOURCE

Mask Disposal Data Collection Sheet

FAMILY FACEMASK USE LOG

Image source:
AleaxmDesign/iStock.com

Keep track of your family's facemask use this week by writing down each member's name. If someone disposes a mask, write the number 1 under that day of the week. At the end of the week, find the sum for each person and then add up the total number of masks disposed on the bottom right.

Name	Mon	Tues	Wed	Thurs	Fri	Sat	Sun	Sum

Total number of masks disposed this week =

▲
Worksheet 1: Mask Disposal Data Collection Sheet

ABOUT THE AUTHORS

Ho-Chieh Lin, a former elementary school teacher in Taiwan and Illinois, is currently a PhD candidate in a STEM education program with a concentration on mathematics education at The Ohio State University. His research focuses on understanding and designing a video-based learning ecology that engages children in authentic discourse for promoting mathematical reasoning and equitable practices.

Joanne Baltazar Vakil holds a PhD in STEM education from The Ohio State University and has been a K–12 educator for 18 years. Currently a research specialist, her interests include Asian Pacific Islander Desi American teacher identity, social media as critical pedagogical tools, learner well-being, and anti-racist initiatives in health sciences education.

- I like knowing people who are like me and different from me, and I treat each person with respect. (Diversity 6)

- I have accurate, respectful words to describe how I am similar to and different from people who share my identities and those who have other identities. (Diversity 7)

- I feel connected to other people and know how to talk, work and play with others even when we are different or when we disagree. (Diversity 9)

MATHEMATICS CONCEPTS

- Adding and subtracting multiples of 10 based on place value and properties of operations.

- Multiplying whole numbers by multiples of 10.

- Understanding and making generalizations about place value; specifically, understanding and justifying that a digit in one place represents 10 times what it represents in the place to its right.

LESSON 5.6 CHALLENGING ABLEIST ASSUMPTIONS IN MATHEMATICS PROBLEMS

Courtney Koestler, Jennifer R. Newton, and Jan McGarry

SOCIAL JUSTICE CONNECTION

This lesson explores the issue of human diversity (i.e., different kinds of bodies), disability, and ableism. This lesson is meant to be launched during or after a typical lesson found in many textbooks that assumes students all have "typical" bodies, such as having 10 fingers, and are able to participate in "typical" ways. Students can use critical literacy skills to examine the mathematics lesson as presented as usual (in many textbooks) as well as resources in their classroom to see how bodies, disability, and ableism are presented. Oftentimes the topic of disability in mainstream classrooms is invisible or explicitly not talked about unless absolutely necessary, and it is important for students to see both children and adults with disabilities represented in their classrooms through empowering ways. Disabilities should be portrayed in ways that avoid deficits and stereotypes and instead accurately describe the disability and/or portray people with disabilities living their lives (whether or not disability is the focus).

DEEP AND RICH MATHEMATICS

This lesson engages students in using their bodies (i.e., their fingers) as a physical representation to support skip-counting groups of 10. At the same time, students will also unpack this common practice to begin a conversation about body diversity.

Resources and Materials

- Chalkboard/whiteboard, chalk/markers, or someplace else to record students' thinking

- Choose among the following books to include in your classroom library:

Picture Books

- *All Are Welcome* by Alexandra Penfold

- *The Bug Girl* by Sophia Spencer

- *Emmanuel's Dream: The True Story of Emmanuel Ofosu Yeboah* by Laurie Ann Thompson and Sean Qualls

- *Hello Goodbye Dog* by Maria Gianferrari

- *I Am Not a Label* by Cerrie Burnell

- *A Kids Book About Disabilities* by Kristine Napper

- *Mama Zooms* by Jane Cowen-Fletcher

- *Rescue and Jessica: A Life-Changing Friendship* by Jessica Kensky and Patrick Downes

- *Terry Fox and Me* by Mary Beth Leatherdale

- *What Happened to You?* by James Catchpole and Karen George

Chapter Books

These may be more appropriate for older grades.

- *Braced* by Alyson Gerber

- *Roll With It* by Jamie Sumner

- *Intersectional Allies: We Make Room for All* by Chelsea Johnson, LaToya Council, and Carolyn Choi

Reference Books for You

- *Critical Literacy Across the K–6 Curriculum* by Vivian Maria Vasquez

Additional Resources

- Article: "How to talk to your kid about disabilities," by Caroline Bologna, *Huffington Post*, March 1, 2021 (https://bit.ly/32Svi68)

- Lesson: Learning for Justice, "Picturing Accessibility: Art, Activism and Physical Disabilities" (https://bit.ly/3oeLVkC)

- Lesson: Learning for Justice, "What Is Ableism?" (https://bit.ly/3oe0kh9)

- Lesson: Learning for Justice, "What Is a Disability?" (https://bit.ly/3lrFFUG)

MATHEMATICS PRACTICES
- Construct viable arguments and critique the reasoning of others.
- Look for and make use of structure.
- Look for and express regularity in repeated reasoning.

LESSON 1 FACILITATION

Introducing Assumptions

Launch (30 minutes)

- A common elementary mathematics activity is to count by tens by counting all the fingers in the classroom. This implicitly makes an ableist assumption that all people have 10 fingers. This lesson engages students in questioning those assumptions and thinking about different mathematical contexts we can use for making tens. Prior to implementing this lesson, we recommend you consider your classroom and how you want to frame the topic. This is important for all classes, but it is especially critical if you have any students who do not have 10 fingers. The *Resources and Materials* section has several links that can be helpful in thinking about how to discuss these topics with your students. If you have a student(s) who does not have 10 fingers, we suggest you discuss the lesson with the family and child to make sure you are approaching it in a way that feels inclusive and supportive to the student(s). While we provide suggestions for some possible approaches to navigating this space, you should adjust based on your context. Say something to students like this:

  I have been thinking a lot about today's activity and wanted to talk to you about it. It is an activity that is in a lot of textbooks because it is usually really good at getting kids to think about patterns in our number system, but I also am wondering about some assumptions it makes about kids and bodies. Let's look at the task and think about some of the hidden assumptions it makes before we start.

TEACHER NOTE

You should adjust what you would say depending on your context. For example, if you have already done work with critical literacy, your students may be familiar with the idea of how authors' assumptions and biases can be analyzed. If not, you may have to discuss it a bit more, perhaps by asking if they know what an assumption is and if they can give any examples. For more information, see Vivian Maria Vasquez's (2016) book, *Critical Literacy Across the K–6 Curriculum*.

- Say to students: *The activity that is usually in textbooks is to "figure out the total number of fingers in our class by counting by tens."* (You may want this displayed on the board as well.) If students need additional support, you may want to explicitly ask:

  + *What does this problem assume about people's bodies? About their fingers?*

  + *How do you know the problem assumes this?*

- Have students share their responses and ask the class: *Does everyone in our world have 10 fingers? Why or why not?* Students may have examples of people in their lives who were born with a different number of fingers or who lost a finger(s) during their lifetime. If not, explain that this could happen. If you have a student who does not have 10 fingers in your class, be mindful of not expecting them to speak as a "representative" for those with a different number of fingers and of not allowing other students to do the same.

- At this point, we offer two possible ways to move forward in the lesson. You may choose to continue with the activity as it is commonly stated (count by tens to find the total number of fingers in the class), or you may ask the class to brainstorm other things that come in tens that you can use for the problem (we especially recommend this second approach if someone(s) in the class does not have 10 fingers). You might tell the class: *The mathematics goal for this problem is to count by tens, but since not everyone has 10 fingers, let's think of some other things that could come in tens that we can count by instead.* The class might suggest ideas like boxes of markers, bags of marbles, or the tens sticks from base-10 blocks. Choose one of these, and prepare one for each person in the class plus several extras before continuing the lesson. Alternatively, you can have the base-10 sticks available and suggest to the class that you use these to represent whatever group of 10 objects the class has agreed on. Give each student one of the tens.

- Tell the class, *We're going to count by tens to see how many ___ we have in our whole class.*

- Have students raise their hands or objects as they say their numbers (10, 20, 30 . . .). You can keep track of the count by listing the numbers, drawing a number line, or using other ways your students suggest to record the numbers.

Explore (30–45 minutes)

- Try it more than once, starting with different students. Ask students to describe what they notice. Ask students questions about what would happen in different scenarios. Here are some examples:

  + *If there are any students absent from class, what would be the number of [objects] if they were present?*

  + *What if the art, music, and physical education teacher joined the class?*

  + *What if we had 35 students in our class?*

  + *What if we counted the whole third grade, which has 78 students?*

- Be sure to entertain multiple strategies for solving the tasks (e.g., counting by tens, adding on by the multiple of 10). As you work, record the problems and solutions (e.g., $18 \times 10 = 180$) in a vertical list off to the side or on chart paper that you can refer back to in the next lesson when the class will look for patterns.

Summarize (20 minutes)

- Briefly summarize the mathematical strategies that students used or ask them to identify any common strategies. Then say that you'll be exploring problems with tens more next time. Next, read one of the suggested children's books. It may be good to have several books to choose from and to begin a study of how disability is explored similarly and differently in the books. We suggest that you continue to have books like those suggested in the *Resources and Materials* in your classroom throughout the school year so that students see these are a part of the classroom library, and not just as part of a lesson on disability.

- While you are reading and/or once you are done reading, have students share questions they have about differences and disabilities.

LESSON 2 FACILITATION

Exploring Place Value

Launch (20 minutes)

- Remind the class of your work last time, both about questioning the hidden assumptions in mathematics problems and your mathematical strategies. Explain that today you will focus more on the mathematical strategies and in a future lesson you will return to exploring hidden assumptions and inclusion in mathematics problems.

- The list of problems from Lesson 1 (e.g., $18 \times 10 = 180$) should be displayed to the side of the board, but do not focus on it yet. Pose some questions like the following to the class: *How many tens does it take to make 780? How do you know?*

- Review the meaning of the place value for a number like 780. What does each digit stand for? Point out that the 7 stands for 7 *hundreds*, and ask them how many *tens* we can think of this as. Point out the connection to the previous problem: 780 stands for 7 hundreds and 8 tens, but it can also be thought of as 78 tens.

- Review the list of problems solved from Lesson 1 (e.g., $18 \times 10 = 180$) and the list of problems solved today. Ask the students if they notice any patterns. You can also ask more explicitly: *What happens when we multiply a number by 10?* Students will likely notice that you "add" a zero to the end of the number. You should question or clarify the use of the word "add" and point out that they don't mean, for example, 18 + 0. You can suggest that another word they can use is "append" or that we might think of this as moving the digits one place to the left (i.e., ones move to the tens place, tens move to the hundreds place, and so on).

- Write this as a class conjecture. If students are not familiar with that term, explain that it's something that you think is true in mathematics, but you need to explore it more to determine if it's always true or not. Tell the class that today they will be working with examples, base-ten blocks, and pictures to try to determine if this is always true and to then prove that it's true.

Explore (20 minutes)

- Students work in pairs or small groups to explore the conjecture. They may want to begin with several examples, then ask them to think about how they can show that it would work with any numbers. Encourage them to think about how they can use base-10 blocks and/or drawings to show what is happening and why that makes this true. Question groups about how they can use what they learned from the warm-up to help them on this problem (e.g., that 780 has 78 tens).

- It may help to have students first think about single-digit numbers (e.g., 7×10) before thinking about double-digit numbers. Circulate among the groups, see what they are thinking, ask probing questions, and make decisions about what explanations to have students share and in what order.

Optional Extensions

It is likely that most students will implicitly stick to whole numbers. Depending on the progress and understanding of the class, you may choose to challenge individual groups or the whole class with different types of numbers such as fractions or decimals. This can be helpful for a conversation about how to modify and/or limit conjectures. The conjecture could be limited to whole numbers, or it could be modified to include decimals by being precise about shifting digits one place value to the left as opposed to appending a zero. The conjecture does not apply to fractions. You can also explore what happens when we *divide* a number by 10.

Summarize (30 minutes)

- Have some or all groups share their thinking about why multiplying by 10 results in appending a zero to the end of the number. Before starting, tell students that when they listen to groups share, they should think of compliments (i.e., something they found valuable about the group's *mathematics*) and questions (i.e., things they didn't understand or want to understand better). The class should engage in a discussion around the different explanations.

- End by highlighting the key strategies that students used and emphasize that mathematicians often spend a long time developing a clear proof that something is true, so it's not something you usually figure out in one day. Depending on the progress made, you may choose to revisit this concept in the future.

- Explain that next time you will return to a focus on the hidden assumptions in mathematics problems.

TAKING ACTION

Remind the class that this lesson began by identifying the hidden assumptions in the common mathematics task: *Figure out the total number of fingers in our class by counting by tens.* Read the book, *Intersectional Allies: We Make Room for All*, by Chelsea Johnson, LaToya Council, and Carolyn Choi. As with any book, you should prepare by prereading *Intersectional Allies*, as there is text in other languages.

Discuss generally what it means to be an ally to others, asking students what they think the word means and asking for examples of allyship. For example, you may ask: *What are the ways that the children in the book acted as allies for their friends? What are ways families supported other families?*

Next, discuss how allyship was framed in the books you used in the previous section. Were there friends, adults, and others that offered supports and accommodations that provided access to people in the books? In what ways?

Taking Action Option 1

Ask students if they know of examples of supports and accommodations that provide access to people with disabilities at their school or in public buildings (e.g., automatic door openers, braille lettering on signs, accessible parking, accessible restrooms).

Ask students to analyze the ways in which the school building is welcoming and safe for different kinds of people, especially for those with different kinds of disabilities. If possible, invite a guest speaker, such as a local disability advocate, to collaborate.

If or when students find issues with accessibility, support them in taking action by communicating via letters or a presentation with building and district administration, school board members, and community members.

Taking Action Option 2

As an ongoing investigation, have students examine ways in which people are portrayed in the books in your classroom, including in your mathematics curricular materials.

For example, you may compare how people with disabilities are represented in *All Are Welcome* by Alexandra Penfold (i.e., where the children just happen to be using a wheelchair or a white cane but not specifically discussed as having disabilities) versus in *Emmanuel's Dream: The True Story of Emmanuel Ofosu Yeboah* by Laurie Ann Thompson and Sean Qualls (i.e., where his true-life story is illuminated about what it was like growing up with a disability).

Students can note places in their textbooks where there are assumptions that everyone in the classrooms is the same, especially in terms of being able-bodied. They may choose to take action by writing letters to different audiences, such as the textbook publishers or district administrators (curriculum coordinators), to describe their findings, let them know how this is not an accurate depiction of people in their classroom and/or world, and give suggestions of ways to make the task or lesson more inclusive. While this activity is ideally student-led, students may need some assistance in developing more inclusive tasks.

COMMUNICATING WITH STAKEHOLDERS

Before teaching this lesson, you should reach out to families, parents/caregivers, and also administrators in your building to provide an overview of the topics included in this lesson (different kinds of bodies, disability, and ableism) and the kinds of the discussions that might emerge. This will help you anticipate ways to be sensitive to and inclusive of the students in your classroom. Any information you receive about specific students or their family members (about differences or disabilities) should stay private, unless they give you explicit permission to share. And, as mentioned earlier in the lesson, take care not to place any student(s) in a position where they have to speak as a "representative" for those who are different or who are disabled.

ABOUT THE AUTHORS

Courtney Koestler is a proud former public school teacher and currently serves as the Director of the OHIO Center for Equity in Mathematics and Science in the Patton College of Education at Ohio University. Their work centers on critical literacy and critical pedagogies in early childhood and elementary education.

Jennifer (Jen) R. Newton began her career as an inclusive early childhood educator in 2000, and has worked across states and settings to promote inclusive practices for students with disabilities. The opportunity to collaborate with Courtney has enabled her to advance anti-racist and anti-ableist work with teacher candidates and teacher education broadly.

Jan McGarry is an elementary teacher in Athens, Ohio. She has had the privilege of working with first and second graders in Appalachia for 20 years. Jan has a passion for fostering inclusive classroom families that center students' voices and encourage connections with the community and current events through the lens of social justice education.

MATHEMATICS LESSONS ON SOCIETY AND SOCIAL MOVEMENTS

CHAPTER 6

Teaching mathematics for social justice (TMSJ) lessons engage students in using and learning mathematics as they better understand social injustices they recognize in their lives. TMSJ engages students in critical analyses of society and empowers them to confront and solve real-world challenges that they face. Sometimes, these actions toward change occur through the building of movements where people are not working in isolation, but rather working together with fellow advocates toward equitable change.

LESSONS

Lesson No.	Lesson Title	Mathematics Focus Areas	Social Justice Topic	Authors
6.1	*"Tu lucha es mi lucha": Mathematics for Movement Building*	• Fraction Concepts and Operations (Decimals)	Inequitable Treatment and Movement Building	Gloria Gallardo and Cathery Yeh
6.2	*Exploring Equitable Pay for Work*	• Fraction Concepts and Operations • Data Concepts and Statistical Thinking	Wage Disparity	Izzy Hendry, Trisha Huynh, and Emma Gargroetzi
6.3	*Modeling Library Funding*	• Fraction Concepts and Operations • Data Concepts and Statistical Thinking • Modeling	Education Justice	Hyunyi Jung and Megan Wickstrom
6.4	*The Value of a School Lunch*	• Fraction Concepts and Operations	Food Insecurity	Rebecca Ellis, Debasmita Basu, Bethany Chan, and Frances K. Harper
6.5	*More Than an Athlete*	• Data Concepts and Statistical Thinking • Whole Number Concepts and Reasoning • Fraction Concepts and Operations	Voting Rights	Evan M. Taylor
6.6	*Your Action Saves Lives: COVID-19 and Systems Thinking*	• Fraction Concepts and Operations (Decimals)	Health Care Justice	Jennifer Park

SOCIAL JUSTICE OUTCOMES

- I know that the way groups of people are treated today, and the way they have been treated in the past, is a part of what makes them who they are. (Diversity 10)

- I know when people are treated unfairly, and I can give examples of prejudiced words, pictures, and rules. (Justice 12)

- I know about the actions of people and groups who have worked throughout history to bring more justice and fairness to the world. (Justice 15)

- I will speak up or do something when I see unfairness, and I will not let others convince me to go along with injustice. (Action 19)

MATHEMATICS CONCEPTS

- Interpret data distributions to answer questions and pose further questions.

- Use the number line for representing fraction (or decimal) magnitudes and operations.

MATHEMATICS PRACTICES

- Model with mathematics.

LESSON 6.1 "TU LUCHA ES MI LUCHA": MATHEMATICS FOR MOVEMENT BUILDING

Gloria Gallardo and Cathery Yeh

SOCIAL JUSTICE CONNECTION

The United Farm Workers of America (UFW) is the nation's first enduring and largest farmworkers union. This lesson will help students connect the inequities faced by farmworkers then to the inequities in labor rights that continue today. Learning about these struggles through the context of mathematics provides students an opportunity to understand the power in using mathematics to analyze (in)justice and build movements for social justice. There is power in numbers and using mathematics to bridge differences.

DEEP AND RICH MATHEMATICS

This lesson was designed to introduce students to decimal concepts and operations. Students learn about decimal concepts and engage in decimal calculations to examine and compare the differential wages of farmworkers and how numbers were used to build solidarity among the different groups of farmworkers.

Resources and Materials

- Book: *Journey for Justice: The Life of Larry Itliong* by Dawn B. Mabalon and Gayle Romasanta

- Video: "AAPI Civil Rights Heroes—Asian Americans Advancing Justice," from the Zinn Education Project (https://bit.ly/32F9r1N) (Scroll down the page to find the video.)

- Article: "Farmworker Wages in California," by Philip Martin and Daniel Costa, *Economic Policy Institute Working Economics Blog*, March 21, 2017 (https://bit.ly/3rMdS5E)

- Article: "Mapping UFW strikes, boycotts, and farm worker actions 1965–1975," by Katie Anastas, Civil Rights and Labor History Consortium (https://bit.ly/3EBYHiO)

- PBS Documentary: *The Farm Worker Movement* (https://bit.ly/2ZIjzWq)

- Mathematics manipulatives (base-10 blocks, fake coins and dollars) for representing wages

- Worksheet 1: *Place Value Chart* (Lesson 2 for decimal understanding and representation)

- Worksheet 2: *Making Sense of the Wages* (Lesson 2 for understanding what workers earned)

- Teacher Resource 1: *Resources for the Gallery Review* (Lesson 3)

LESSON 1 FACILITATION

History of the Farmworkers' Movement

Launch (20–30 minutes)

- Begin the lesson by asking students to *draw a farmworker and then on the back of their drawings write three adjectives they think describe their "worker."*

- Ask students to place all their drawings for display. Lead the class in discussing their observations, including commonalities and differences among the drawings. Commonalities and differences can include common adjectives, race and ethnicity of the farmworker, and other observations about the farmworker.

- Read aloud and reflect on the list of adjectives. Sort them into positive/affirming terms and negative/denigrating terms. Ask students to discuss how this naming influences how they themselves perceive farmworkers and point out the importance of self-identifying one's position or bias when examining social justice issues.

- Ask:

  + *What does "counter-narrative" mean?*

  + *How can we unpack that word?*

  + *What are examples of counter-narratives?*

- Use examples to discuss collectively that **counter-narratives** offer alternative perspectives to an existing story. The perspectives represented by counter-narratives are often not explored or known by many people. Ask students questions such as these:

  + *Why is it important to hear or learn about different perspectives of the same story?*

  + *What can we learn from different perspectives?*

Counter-narratives are an important part of teaching for social justice. They offer historical accounts and interpretations that question both dominant white and single (one-perspective) narratives.

+ *Can anyone think of a counter-narrative that others may not know about?*

Student responses will vary depending on grade level and teacher scaffolding, but might include those seen in Figure 1.

Figure 1. Possible Student Responses on Learning About Different Perspectives

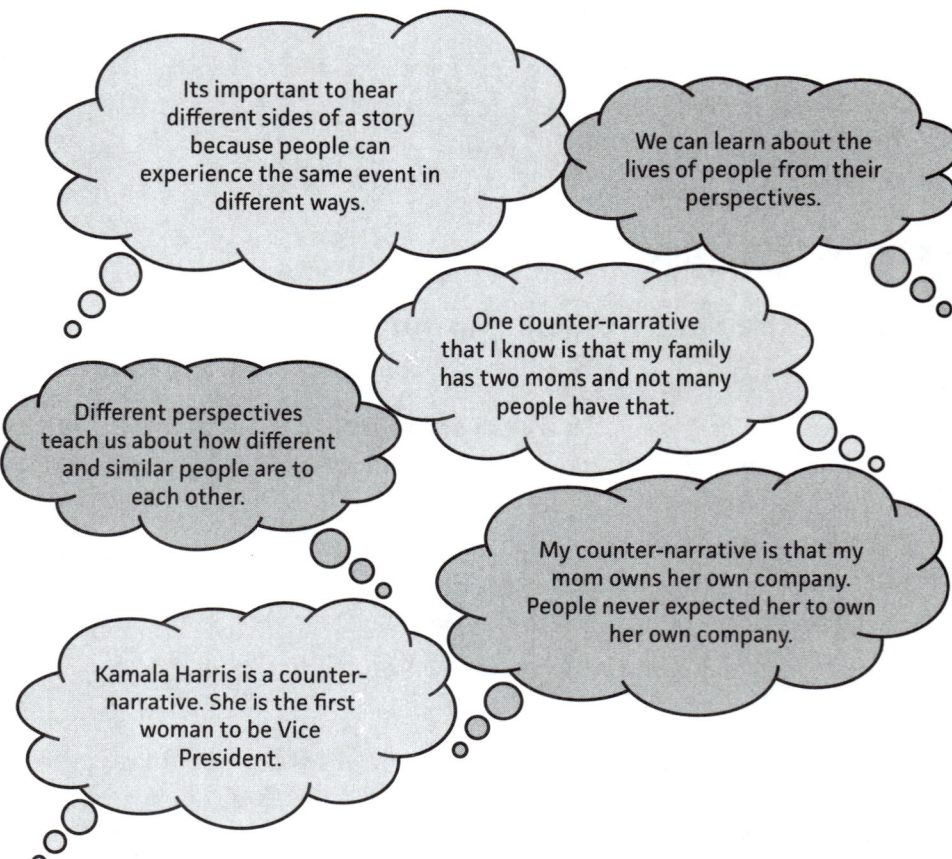

- Show students the UFW website (https://ufw.org/). Tell students that this lesson series focuses on the counter-narrative of the farmworkers' movement from the 1960s from the perspective of a labor union organizer named Larry Itliong and the Filipinx migrant workers who fought alongside him.

- Ask: *What do you know about Larry Itliong? Do you have family members that are a part of UFW?*

- These questions allow us to assess student background knowledge of Larry Itliong and UFW as well as open up the invitation to bring in caretakers, family members, and local UFW branches and labor unions to be a part of the lesson.

- Student responses will vary depending on grade level and teacher scaffolding, but might include these:

    + *My family is Filipino, but we don't know about Larry Itliong.*

    + *Larry Itliong was a Filipino farmworker. My sister told me about him.*

    + *I know about UFW. Cesar Chavez started UFW.*

- Introduce the key terms of the lesson: *migrant workers, farmworkers, welga/huelga, boycott, union, wage, decimals, place value, inequity, inequality, cost of living,* and so on. As the lesson progresses, students can create a concept map on the farmworkers' movement to provide a visual of how these terms/ideas are connected. Perhaps students can ask their parents/caregivers their remembrances of this movement to increase their collective knowledge and consciousness. Refer to the concept map as a class daily as the lessons progress.

- Depending on class demographics and school settings, some students may possess knowledge of the farmworkers' movement or make connections to what they know about farmworkers or labor history and rural regions. For example, a key standard in Californian history and social studies examines the differences between the state's four natural regions: desert, mountain (Sierra Nevada), coastal, and valley (Great Valley). Students will have learned that all agricultural production occurs in the Central Valley Region of California. They will also have learned that the key labor force of this region is farmworkers.

- Build from the understanding and connections that students have about the farmworkers' movement. Meet them where they are and validate their knowledge and understanding because it is the narrative that they know.

Explore (20–25 minutes)

- Begin reading *Journey for Justice: The Life of Larry Itliong* to the class. This book is the heart of this lesson, as it introduces an often-forgotten leader and a key part of the history that led to the organization of UFW.

- As a class while you read, make a large timeline (one side of the classroom wall) of Larry's journey as a labor organizer and his wages (up to page 25 of the book). This will be a visual representation of Itliong's journey for justice and the significant moments throughout his life of service.

- Stop at pivotal moments where the author describes how much Larry and other Filipinx migrant workers were paid for their labor. Write down the different hourly rates Larry earned as he traveled from farm to farm. Pass out place value mats, coins, and base-10 blocks. Allow students to represent the hourly wages using the different physical representations. Here are some examples:

+ *When Larry worked as a janitor at the Frye Lettuce Farm, he was paid 12 cents per hour.*

+ *Later in 1939, he strikes with other workers on the asparagus fields to protest their wages of 10 cents per hour.*

+ *The growers threatened to cut their wages by 10%. (Ask the class how much the workers would be paid then, if they were paid 10 cents per hour but were going to lose 10% of their wages.)*

+ *Lastly, by 1965 Larry was actively helping Filipinx workers organize for fair wages. Filipinx workers in Coachella Valley were paid $1.25 an hour.*

Summarize (10–15 minutes)

Show students the 2-minute video interview of Larry Itliong's son discussing Larry's life contributions. Go to the Zinn Education Project (https://bit.ly/32F9r1N) and choose the second video, "AAPI Civil Rights Heroes—Asian Americans Advancing Justice."

Larry's story is an example of a counter-narrative—the untold stories of the struggles and contributions of people of Color to U.S. history, arts, and civil rights. Ask students to honor an unsung hero in their life and what that person does to build community and to advocate for justice for all.

LESSON 2 FACILITATION

Events That Led to Unity Between the Filipinx and Latinx Workers

Launch (15 minutes)

- Start the lesson by showing the first few minutes of the PBS documentary, *The Farm Worker Movement* (https://bit.ly/2ZIjzWq).

- Show the photo of the mural of Philip Vera Cruz and Larry Itliong and read the following narrative provided by the Zinn Education Project (n.d.) on the Delano Grape Strike (https://bit.ly/32F9r1N):

On September 8, 1965, Larry Itliong, a member of the Agricultural Workers Organizing Committee, called a strike against the Delano, California, grape growers in order to demand salaries equivalent to the federal minimum wage and the right to form their own union.

Sept. 8, 1965, at the Filipino Hall at 1457 Glenwood St. in Delano, the Filipino members of AWOC held a mass meeting to discuss and decide whether to strike or to accept the reduced wages proposed by the growers.

The decision was "to strike" and it became one of the most significant and famous decisions ever made in the entire history of the farmworkers struggles in California. It was like an incendiary bomb, exploding out the strike message to the workers in the vineyards, telling them to have sit-ins in the labor camps, and set up picket lines at every grower's ranch. . . – Philip Vera Cruz

More than one thousand Filipinx workers walked out of their grape farms to picket, however, farmers replaced them with Mexican workers.

- Share with students that Filipinx and Latinx laborers worked in separate fields and lived in separate camps. Have students compare the differential wages between workers, with some paid at $1.40 per hour but Filipinx workers only given $1.25 per hour. Encourage students to use manipulatives and other tools to represent each quantity. You could also have students calculate the wages daily, weekly, or monthly and in relation to information about the cost of bread, a gallon of gas, or a small home during that time. Ask: *Could a worker support their family on those wages?*

- Choral counting by $1.25 (see Figure 2) can support students' understanding and flexibility with decimal concepts and operations. Choral counting is an activity in which the teacher leads students in counting aloud together by a given number. Different numbers lend themselves well to surfacing different mathematical ideas.

Figure 2. Choral Counting by $1.25

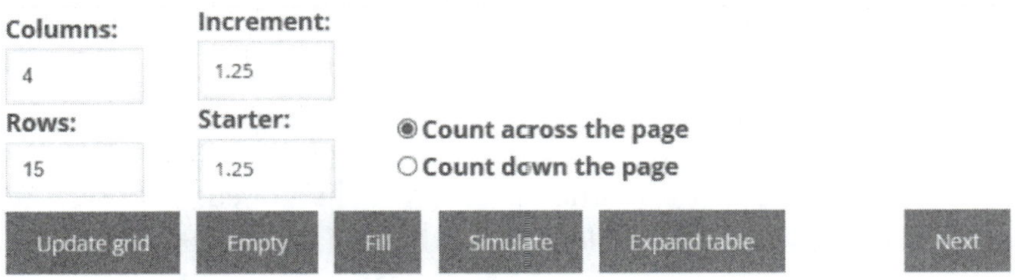

Columns: 4

Increment: 1.25

Rows: 15

Starter: 1.25

◉ Count across the page
○ Count down the page

[Update grid] [Empty] [Fill] [Simulate] [Expand table]   [Next]

1.25	2.50	3.75	5.00
6.25	7.50	8.75	10.00
11.25	12.50	13.75	15.00
16.25	17.50	18.75	20.00

In the early 1960s, manongs, **Filipinx** farmworkers, were paid $1.25 per hour of labor. On average, farmworkers were expected to work __12__ hours a day, 6 days a week. How much did **the manongs** earn in one day? The manongs had little money and relied on low-income housing in which rent was **$50 a month**.

How much would the manongs make in a month? How much money would the manongs have left over after paying rent?

Explore (45 minutes)

- Divide students into heterogeneous groups of three to four with access to place value charts, base-10 blocks, play money and coins, and calculators. Use the following prompts:

  + *Would the farmworkers be able to make enough to cover their living costs?*

  + *What is the impact of the differential wage per day, per week, and per month compared to the cost of living during 1965?*

- Investigate the wages of those workers and create a graph. Illustrate the wages as decimals and have students calculate the differences. Show students how to calculate the percentage difference between wages (division will result in a decimal that students then convert to a percentage).

- If decimals are a new concept for students, give students time to represent the different values of $1.25 and $1.40 using play money and coins, base-10 blocks (with the hundred block representing $1.00), and place value charts (see Worksheet 1 in the *Resources and Materials*) so they develop comfort with and understanding of decimal concepts before calculating the impact of differential wages. Encourage students to use multiple representations (e.g., play money and coins, place value charts, words, and numbers) to support students' reasoning, explanations, and justifications.

- Monitor students' progress throughout the activity. Draw student attention to the extreme working conditions. At the time, many farmworkers were living in poverty without enough food to eat or other basic necessities. They lacked access to fresh water while working in the fields, some lacked sanitary facilities for human waste, and some were present in the fields as crop dusters dropped pesticides on the crops. Another mathematics connection would be to have the students (using a chart of current prices for rent, utilities, gas, food, etc.) calculate living costs and to see how difficult it would be to survive on the wage being paid (Worksheet 2, *Making Sense of the Wages*).

Summarize (10 minutes)

- Continue reading *Journey for Justice: The Life of Larry Itliong* from page 25 to the end of the story. Throughout the read-aloud, have students add on to the classroom timeline of the boycott strike. Draw students' attention to how the strike led to a movement that drew unprecedented support from outside the Central Valley, from other unions, church activists, students, and civil rights groups, leading to the 300-mile march, or peregrination, from Delano to Sacramento. The strike placed the farmworkers' plight before the conscience of the people, leading to boycotts of table grapes across North America. Millions stopped eating grapes. By 1970, the grape boycott ended when table grape growers at long last signed their first union contracts, granting farmworkers better pay, benefits, and protections. The story of the nonviolent strike, protest, and boycott is rich with numbers.

Stop students during moments of reading to grasp the magnitude of the numbers (e.g., *How far is 300 miles? How many days would it take to walk from Delano to Sacramento? How long was the protest?*)

- Recognize the power of building movements. After the reading, have students name out the many individuals that took part in the strike and protest that led to negotiation: Dolores Huerta, Cesar Chavez, Larry Itliong, Philip Vera Cruz, other labor unions, church activists, students, other minority groups, civil rights groups, and families that boycotted table grapes.

- Cesar Chavez believed in nonviolent protest: "[T]he whole essence of nonviolent action is getting a lot of people involved, vast numbers doing little things." Ask students to give other examples in history (past and present) in which social change occurred through collective action.

TAKING ACTION

Start with a gallery review that includes artifacts, images, data, and videos (migrant farmworker children) about the farmworkers movement from the past and present (see Teacher Resource 1: *Resources for the Gallery Review*). One of our guiding questions is *How does looking at the past help us make sense of now?* Students may reference current worker conditions from farms, factories, or construction sites depending on their regional location and the type of labor prevalent in their areas. This part of the lesson can be extended to bring in local labor leaders or a student family member in this line of work that can co-teach a lesson or to share their experience.

Students will move around the classroom, observing these different images. Students will write down what they notice, what they wonder about, and how these photographs and data make them feel. Their noticings and wonderings will be applied later in their action plan.

After the gallery review, students will share in small groups what they noticed, what they wondered about, how they felt, and what personal actions are inspired by the photos. The purpose of this activity is to help students connect the experiences of farmworkers in the 1960s and 1970s with the experiences of farmworkers today.

While this movement happened a number of years ago, how much has really changed? Ask students to consider the lives and conditions of migrant and labor workers today. Students can learn about current issues, advocacy, and activism of the United Farm Workers from their website (https://ufw.org/take-action/). Teachers can also reach out to the local branch of UFW or other local labor organizations to examine their workplace safety, working conditions, and wages. Investigate the labor concerns shared and ask students to collect, analyze, and share data with their local congressman, county supervisor, or city council member as well as corporations and community leaders to push for change.

In an effort to use more inclusive language, we use the term *gallery review* for what is commonly referred to as "gallery walk."

(Continued)

(Continued)

For example, the UFW is organizing for overtime protection. Farmworkers have been excluded from the federal Fair Labor Standards Act's overtime law since it was passed in 1938. California and Washington are the only states where overtime equity for farmworkers exists.

Students may want to examine the differential salary between workers with overtime pay and regular pay. How much is overtime pay? This information varies by state. For example, California overtime laws require nonexempt employees to earn one-and-a-half times their regular rate of pay when they work

- More than 8 hours in a workday,
- More than 40 hours in a workweek, or
- More than 6 consecutive days in a workweek.

Employers also must pay double-time for nonexempt employees working *more than*

- 12 hours in a workday, or
- 8 hours on the seventh consecutive day in a workweek. (California Labor Code Section 510; California Legislative Information, 2000)

The farmworkers movement was founded on the belief that workers deserve to earn wages they can live on, have health benefits, work in conditions that are safe and clean, and have their humanity and dignity preserved. This movement was about the power of collective efforts, not just individuals. Although we learned about the efforts and influence of specific individuals in this movement, we need to remember that movements are created by groups of people working toward justice.

The farmworkers movement did not end when the Delano grape growers allowed workers to form a union and gave them better wages and some benefits. The farmworkers movement continues today and their cause is still important. Discuss the following in small groups and/or as a whole class: *Why are farmworkers (or other laborers) still paid low wages? What can consumers and people do to support the rights of farmworkers (other workers)? How can consumers make a difference?*

COMMUNICATING WITH STAKEHOLDERS

You should communicate the social justice and mathematics goals of the lesson to parents/caretakers via email, phone calls, and/or the classroom webpage and reach out to caretakers, grade-level teachers, and local branches of UFW and other local labor unions to see if they'd like to be involved in the lesson. You should welcome administrators to plan, observe, and join in the lesson implementation to support and encourage schoolwide implementation of social justice curriculum across content areas.

ONLINE RESOURCES

 Available for download at **resources.corwin/TMSJ-UpperElementary**

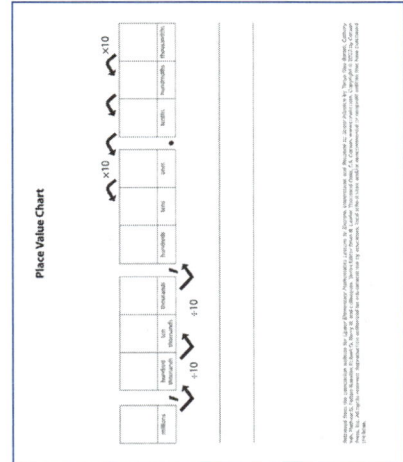

▲ *Worksheet 1: Place Value Chart*

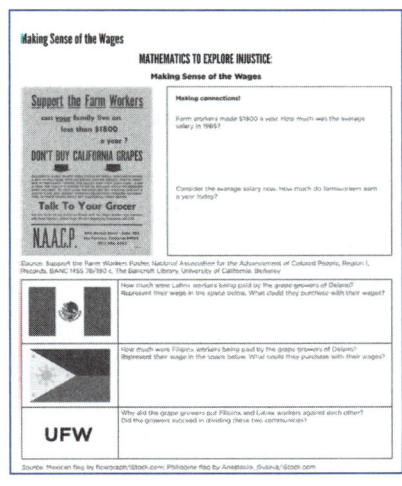

▲ *Worksheet 2: Making Sense of the Wages*

▲ *Teacher Resource 1: Resources for the Gallery Review*

ABOUT THE AUTHORS

Gloria Gallardo is a social justice bilingual educator, woman of Color, proud daughter of Mexican immigrants, intersectional feminist, and aspiring writer. She is a graduate of the Master of Arts in Teaching Program at Chapman University. As a social justice educator, she strives to give power back to her students, and she strives to do so by empowering them to see themselves for all their possibilities and potential.

Cathery Yeh is a parent, bilingual educator, and mama-scholar at the Universiity of Texas at Austin and co-founder of #miseducAsian. Her teaching and research examine the intersections of race, ethnicity, gender, language, and dis/ability identities in K–12 educational contexts and culturally relevant, justice forward, and asset-oriented approaches in K–12 pedagogy.

- I know that the way groups of people are treated today, and the way they have been treated in the past, is a part of what makes them who they are. (Diversity 10)

- I know when people are treated unfairly, and I can give examples of prejudiced words, pictures and rules. (Justice 12)

- I know that words, behaviors, rules and laws that treat people unfairly based on their group identities cause real harm. (Justice 13)

- I will speak up or do something when I see unfairness, and I will not let others convince me to go along with injustice. (Action 19)

MATHEMATICS CONCEPTS

- Focus on real-world contexts for understanding fraction operations conceptually.

- Create data displays to organize, analyze, and communicate data information.

- Interpret data distributions to answer questions and pose further questions.

LESSON 6.2 EXPLORING EQUITABLE PAY FOR WORK

Izzy Hendry, Trisha Huynh, and Emma Gargroetzi

SOCIAL JUSTICE CONNECTION

The social injustice topic of the lesson is the wage gap—the phenomenon of unequal pay based on race and sex. In the United States, women and people of Color are consistently paid less than white men. This phenomenon is often described using the cents that a Latinx woman (for example) is paid for each dollar paid to a white man. Other comparisons are also possible, of course. By opening and closing the lesson with connections to social movements that challenge this injustice, students learn that challenging injustice they face is part of what makes people who they are.

DEEP AND RICH MATHEMATICS

Students use graphing and fractions to investigate wage disparities between people based on their race and gender. Attention can be given to fraction equivalence with a focus on benchmark fractions (thirds, quarters, fifths, tenths) as well as to relationships between fractions and decimals. Students use multiple representations (graphs and fractions) to analyze real-world data that impact their own lives and draw conclusions about the meaning and implications of the data. They also consider which representations might be used to best communicate this information with others.

Resources and Materials

- Worksheet 1: *Data Suggesting Wage Inequalities*

- Graph paper

- Assorted images (three to five) of protests and organizing about the wage gap (teacher choice)

- Poster paper

- Article: "Gender pay gap in U.S. held steady in 2020," by Amanda Barroso and Anna Brown, *Pew Research Center* (https://pewrsr.ch/3y3K6dJ)

- Article: "Quick facts about the gender wage gap," *Center for American Progress* (https://ampr.gs/3Ev42IN)

- Article: Survey data from "The state of the gender pay gap in 2021," *Payscale* (https://bit.ly/3lMLMn0)

Optional
- Book: *Joelito's Big Decision/La Fan Decision do Joelito* by Ann Berlak

- Article: "U.S. women's soccer won 4 World Cups. Now can they score equal pay?" by Gretchen Frazee and Yasmeen Alamiri, *PBS News Hour*, July 8, 2019 (https://to.pbs.org/3Ivy1m7)

MATHEMATICS PRACTICES
- Reason abstractly and quantitatively.
- Construct viable arguments and critique the reasoning of others.

LESSON 1 FACILITATION

Introducing the Wage Gap as a Social Justice Issue

Launch (25–35 minutes)
- Open the lesson by inviting students to name social movements or issues of equity that have been discussed previously in your classroom or that they are aware of from their own lived experiences. Discuss with students the similarities and differences across the examples shared and pinpoint what injustice is highlighted in each example.

- Also, launch the lesson in connection with a story or current event that can provide context for your students. Here are two possible options for launching related to fair pay:

  + Engage in a read-aloud with *Joelito's Big Decision/La Fan Decision do Joelito* by Ann Berlak (2015). In this story, Joelito has to decide whether or not to cross a picket line where workers at a burger joint he likes are on strike for fair pay.

  + Engage in a read-aloud with the article, "U.S. women's soccer won 4 World Cups. Now can they score equal pay?" from *PBS News Hour* (https://to.pbs.org/3Ivy1m7).

- Ask students if they have heard the word *wage* before. Invite students to share where they have heard this word and what it might mean. Introduce how "wage" is being used in this lesson: a wage is how much someone earns for their work.

- Ask students what they know about wage gaps or differences in wages for women or people of Color as compared to white men. This supports you in learning about what students already understand about the injustice of wage inequality along certain demographic lines.

- Tell students: *Today we are going to look at some data and try to figure out what it tells us about different people's wages. As we examine the data, we will consider what patterns we notice, whether what we see is fair, and additional questions we might explore to learn more about this issue.*

Explore (30 minutes)

- Have students work with a partner.
- Show students the wage data graph from Figure 1.

Figure 1. The Gender Wage Gap

The gender wage gap is more significant for most women of color
Comparing 2018 median earnings of full-time, year-round workers by race/ethinicity and sex

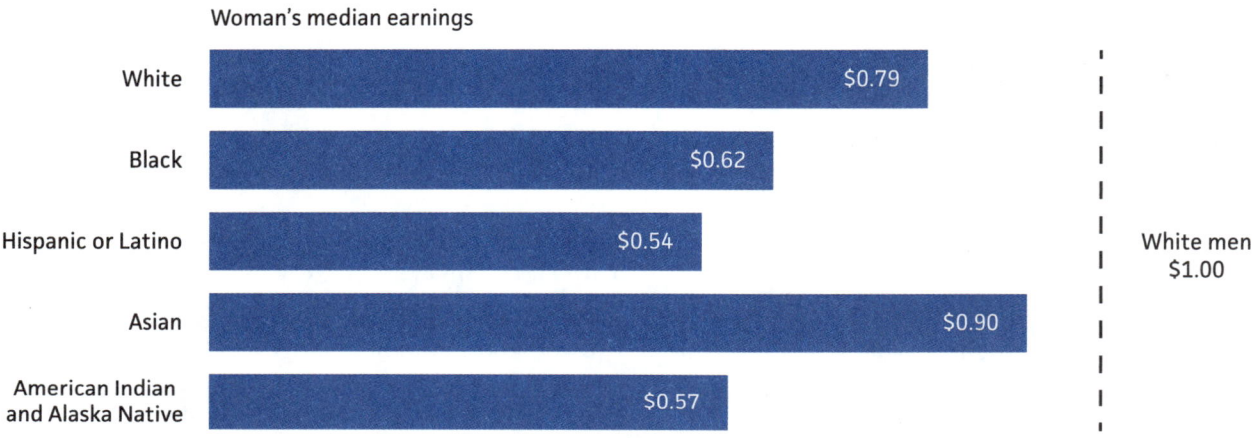

Notes: The gender wage gap is calculated by finding the ratio of women's and men's median earnings for full-time, year-round workers and then taking the difference. People who have identified their ethnicity as Hispanic or Latino may be of any race.

Sources: For all groups except American Indian and Alaska Native women, the Center for American Progress calculated the gender wage gap using data from U.S. Census Bureau, "Current Population Survey: PINC-05. Work Experience-People 15 Years Old and Over, by Total Money Earnings, Age, Race, Hispanic Origin, Sex, and Disability Status: 2018," available at https://www.census.gov/data/tables/time-series/demo/in-come-poverty/cps-pinc/pinc-05.html (last accessed March 2020). Specific tables used are on file with the author.
CAP calculated the gender wage gap for American Indian and Alaska Native women using U.S. Census Bureau, "Table B20017C: American Indian and Alaska Native alone population, non-Hispanic or Latino population 16-years and over with earnings in the past 12 months, 2018 American Community Survey (ACS) 1-Year Estimates," available at https://www.census.gov/programs-surveys/acs/ (last accessed March 2020); U.S. Census Bureau, "Table B20017H: White alone, non-Hispanic or Latino population 16-years and over with earnings in the past 12 months, 2018 American Community Survey (ACS) 1-Year Estimates," available at https://www.census.gov/programs-surveys/acs/ (last accessed March 2020).

Source: This material was created by the Center for American progress (www.americanprogress.org)

- Ask students to turn and talk to their partner about the data presented.
  + *What do you notice?*
  + *What do you wonder?*

- Possible student responses may include the following:

    + *The pay gap is more significant for women of Color.*

    + *This data is from 2018. Is it the same now as it was then?*

    + *There may be differences within a race because not all people of one race are treated the same way.*

- Invite a few students to share from the turn and talk. Document on chart paper wonderings students have:

    + *Why is this happening?*

    + *Is it the same now as it was a long time ago?*

- The following questions may guide sharing and discussion:

    + *Why do we use $1.00 as our base versus an annual salary?*

    + *Why are we comparing all of the categories to one category (a white male)?*

    + *Why don't they split up the data by particular jobs?*

TEACHER NOTE

The choice to do this calculation across all industries is purposeful, as noted by the Center for American Progress ("Quick facts about the gender wage gap," https://ampr.gs/3Ev42IN). Calculating in this way supports experts in identifying the various factors that relate to the wage gap, such as differences across industries (i.e., so-called "women's jobs") and differences in years of experience and hours worked (usually related to women more often accommodating care giving and to discrimination, particularly of women of Color).

- Build off of your student responses to make sure the whole class understands that the data in this graph reflect the ratio of median annual earnings for women working full-time, year-round, to those of their white male counterparts. Discuss students' understandings of the terms *ratio*, *median*, and *annual earnings* in the context of the data (e.g., *What do the "cents" mean? How does the $1.00 base relate to the idea of annual earnings?*). This concept of cents on the dollar may be counterintuitive to students and will likely require further engagement across the lesson to make sense of what this graph is communicating about social and economic injustice.

- Say to students: *Let's consider this data in a different way to further support our understanding of this situation. Remember, when we solve problems, we use multiple strategies and representations to understand new ideas.*

- Use Figure 2 to discuss with students an example scenario: Imagine Mark and Maria both work full-time as business executives. When Mark earns $1.00, Maria earns $0.60.

  + *How does this relate to the graph we looked at?*

  + *How could you talk about Maria's pay as a fraction of Mark's?*

  + *How does the picture support your idea?*

  + *When Mark makes $2.00, how much will Maria make?*

  + *How do you think fractions might help us to understand data?*

- Ask students to represent the data in the graph in Figure 1 with fractions, in writing or models (drawings or with manipulatives). Have available manipulatives such as blank paper, coins, and base-10 blocks or centimeter cubes.

Figure 2. Mark and Maria's Wages

Mark, $1.00   Maria, $0.60

What fraction of Mark's wage does Maria earn?

Summarize (10 minutes)

- Engage students in a thinking routine, either in pairs or in table groups, where students discuss the following questions:

  + *What did you do to make sense of the data?*

  + *How did this data make you feel?*

  + *What did this data make you wonder?*

- Invite pairs or table groups to share in a whole-class discussion. Document wondering questions in the same place you started documenting wonderings during the launch. Repeat feelings back to the student who shared and to the class. For example, if Jojo says they feel confused, you might respond with one of the following:

  + *Yes, this can make us feel confused. Thank you for sharing.*

  + *Yes, this can make us feel confused. Can you say more about what is confusing for you?*

LESSON 2 FACILITATION

Exploring the Wage Gap

Launch (5–10 minutes)

- Ask students to turn and talk with a partner about what they remember from yesterday's work. Elicit responses from the students.

- Point to your list that students started to generate yesterday. Say: *We have generated many wonderings. What questions did we answer? What did you use and do to answer those questions?*

Explore (20–30 minutes)

- Have students work in small groups of three to four students in each group.

- Share the data table in Worksheet 1 (*Data Suggesting Wage Inequalities*). This table looks at women's earnings compared to their white male counterparts' earnings over time.

- Ask students to talk in their small groups about the data table in Worksheet 1. Ask: *What do you notice? What do you wonder?* Invite students to share their thoughts with the whole class. Focus on making sure all students in the class understand what the data in the table represent.

- Discuss with students that these are the classifications used by this particular organization. Ask students to turn and talk with their table group about how they self-identify and how they might define these categories (i.e., Asian, Black). Invite students to share their thoughts with the whole class.

- Tell students: *One way we might further explore this data is by representing it in graph form. It provides us with a visual way of thinking about the data. Talk in your small groups about the following:*

    + *How might you represent this data with a graph?*

    + *How could this data be represented using fractions?*

    + *What information should someone looking at the graph be able to learn?*

TEACHER REFLECTION

Children may have diverse and exciting ideas about how to represent data that are not aligned with the canonical representation of bivariate data on a coordinate plane. Prior to introducing the coordinate plane in this lesson as a way to plot the data provided, we encourage doing prior work with students to explore bivariate data, encouraging them to share and discuss diverse forms of data and how they might collect and represent them.

—Emma Gargroetzi

- Provide student groups with graph paper and have them start to create graphs from the data.

- As students work, monitor how they are representing different categories on their graphs. If they have not seemed to think about this, ask: *How might you represent these different categories on your graph?* Note that students may suggest using different colors to represent the different categories. Acknowledge that this is a good idea—we do want to represent the different categories in the graph. Also ask students how they might help someone who is color blind to understand the data (i.e., using different types of lines rather than different colors).

- You can also confer with student groups to facilitate mathematical and social connections, asking questions such as these:

    + *Do you see any patterns?*

    + *How are the numbers changing over time?*

    + *How did you decide to place these data points? (Focus on those that are in between the benchmark numbers, such as 55 cents.)*

- If students are struggling with the task or with collaboration on the task, you might set a norm prior to beginning the task that any question asked by the group must be a question that all students have (in other words, they need to check with each person in their group to make sure they all have that same question as opposed to working on their own without consulting group members).

- Ask questions such as these:

    + *Where can we put the paper and markers so that all people can see the paper and use the markers?*

    + *What ideas are you having about how to start?*

    + *What makes this one* (the nonbenchmark number) *different from some of the other data points you already represented?*

Optional Additional Exploration

- Inform students: *The way we have explored this social justice issue thus far—in terms of dollars and cents—helps us see some things, but it may also mask the true impact of this inequity on working women and their families.*

    In your small groups, explore this problem:

    + *A woman working full-time, year-round earned $10,194 less than her male counterpart, on average, in 2018. If this wage gap were to remain unchanged, how much less would she earn than a man over the course of a 40-year career?* (Data taken from the Center for American Progress, https://ampr.gs/3Ev42IN)

- Repeat this calculation for each group's data in the following table:

Group	Individual Earnings Gap in 2018	Individual 40-Year Earnings Gap
All women	$10,194	
White women	$13,186	
Black women	$23,540	
Hispanic or Latina/o Women	$28,036	
Asian Women	$6,007	
American Indian and/or Alaska Native Women	$25,884	

Summarize (10 minutes)

- Engage students in a thinking routine, either in pairs or in table groups, where students discuss the following questions:

    + *What did you learn today?*

    + *What challenges did you face?*

> + *How did this data make you feel?*
>
> + *What did this data make you wonder?*

- Invite pairs or table groups to share in a whole-class discussion.

LESSON 3 FACILITATION
Moving to Action

Launch (10 minutes)

- Have students turn and talk to their partner about some things they remember from the previous two lessons.

- Invite students to share. Use their ideas to begin to summarize and capture big ideas and things that they have learned.

- Explain: *Today we will reflect on our learning together, try to connect what we have learned to other learning we are doing together, and make a plan for what to do next.*

Explore (15 minutes)

- Place a large sheet of poster paper on each classroom table or around the classroom on the walls. In the middle of the poster paper, glue or tape down either a question alone or an image with a question. Here are some examples:

  > + *What surprised you?*
  >
  > + *What did you learn?*
  >
  > + *What questions do you still have?*
  >
  > + *How could this image connect to what we learned about the wage gap?*

Images

The goal is for students to use images to see examples of how people have challenged injustices in the past. Therefore, the images you use could come from mathematics, ELA, science, and/or social studies units that students have engaged with before in class that provide examples of forms of protest and, if possible, examples of people fighting for equal pay.

You might also draw on images from the list in Figure 3.

Figure 3. Examples of Forms of Protest for Equal Pay

Books

- A page from *Joelito's Big Decision* book

Images and Websites

March on Washington. Source: https://bit.ly/3DqrRSo

Source: National Archives Identifier

- President John F. Kennedy signing the Equal Pay Act (https://bit.ly/3GjMIa0)

- Encountering Racism With Courage and Protest: Protest for Equal Pay (https://bit.ly/3y5tbaC)

- Equal Pay for Native Women (go to http://www.equalpaytoday.org and find the link for Native Women)

- Latina Equal Pay Day (https://www.latinaequalpay.org/about)

Articles

- "The fight for equal pay . . . 40 years on," by Jo Revill, *The Guardian*, May 31, 2008 (https://bit.ly/3oxupYY)

- "1970 Women's Strike for Equality" by Linda Napikoski, *Thought.Co*, February 24, 2019 (https://bit.ly/3rJCQT9)

- "Equal Pay Day: Women rally against wage gap, workplace discrimination," *NBC News*, April 4, 2017 (https://nbcnews.to/3EPHnqZ)

- "Equal Pay Day for Asian American and Pacific Islander Women," from the Feminist Majority Foundation (https://bit.ly/3IsvA3T)

- "Equal Pay for Black Women," from National Women's Law Center (https://bit.ly/3ozVevE)

- "Those with disabilities earn 37% less on average; Gap even wider in some states," American Institutes for Research (https://bit.ly/3DuPuaF)

- "The gay and transgender pay gap," by Center for American Progress (https://ampr.gs/3EEVOOk)

- "Before Team USA Women's Hockey won Olympic gold, they won equality off the ice," by Alix Langone, *Money*, February 18, 2018 (https://bit.ly/33bSlJl)

- "The root of WNBA players' argument about gender wage gap," by Adam Grosbard, *The Atlanta Journal Constitution*, June 28, 2018 (https://bit.ly/3IrB3YK)

- Point out to students that around the room there are posters with a question or sometimes with an image and a question in the middle of each poster. Explain that there will be two rounds of poster visiting. In the first round, they will look at what is in the middle of the poster and write their ideas onto the poster in response to the question and/or image in the middle of the poster. The second time around, they will read what their classmates have written and can respond in writing if they desire.

- This is initially a silent and solo activity (although students may circulate with a small group). Provide about 2 minutes at each poster. You should signal students when they should move from one poster to the next and which poster to visit next.

- As students respond to the posters, note mathematical ideas and ideas about social injustice as well as any misconceptions or stereotypes that will be important to address as a class.

- Figure 4 includes some responses from a fourth-grade, dual language class in San Francisco that engaged in this activity. The student spelling is original. Translation is by Izzy Hendry.

Figure 4. Student Responses on the Equal Pay Lesson

What did you learn?

"Yo ampredi [aprendí] como ase uno graphic." *I learned how to make a graph.* —M

"Yo tambien aprendi hacer mate en diferente manera." *I also learned to do math in a different way.* —M

"Yo aprendi que no tengo que hacerlo del mismo color y que no tenemos que hacerlo lo mismo que lo anteriores." [The plot points for different subgroups] *I learned I shouldn't make the lines the same color and that we shouldn't make them the same as the previous one.* —L

"Yo estoy de acuerdo contigo porque despues si usas el mismo color te confundes." *I agree with you because later if you use the same color you get confused.* —B

What questions do you still have?

"Una pregunta que aun tengo es porque le pagan a un hombre blanco un $1 dolar si las otras personas tienen el mismo trabajo y los pagan diferente?" *A question I still have is why they pay white men $1 if the other people have the same job and they pay them differently?* —H

"Mi pregunta es proque a los hombres hispanic les pagan mas y alas mujeres hispanic les dan menos?" *My question is why Hispanic men get paid more and Hispanic women get paid less?* —A

"Yo estoy desacuerdo porque los hombres asen mas trabajo." *I disagree because men do more work.* —O

"Tal vez porque el jefe del trabajo no les quiere pagar el dinero (lo que se meresca)." *Maybe it's because the boss doesn't want to pay them the same salary (that they deserve).* —H

"Porque las mujeres hispanas ganaron menos dinero?" *Why do Hispanic women earn less money?* —M

"Yo creo porque no son white men y porque son mujeres." *I think it's because they aren't white men and they are women.* —L

"Yo estoy acuerdo." *I agree.* —R

What surprised you?

"A mi me sorprendio como a la muejr le subieron el pago porque primero le pagaron 60 c y luego 80c." *I was surprised that women's pay went up because first they paid them 60 cents and later 80 cents.* —A

Summarize (15 minutes)

- Bring the class back together (in a circle on the rug or however you generally bring the class together in your setting). Spread posters out in the middle of the class circle. Ask students to observe each other's responses that they might not have gotten to read yet. You might have students quietly review the posters for a few minutes before settling down into a circle to talk.

- Ask students what additional questions they have based on what they have learned. Some extension questions that might arise include the following:

    + *Does women (or another group) getting paid less only affect women? How might it affect their family and community, too?*

    + *In what ways did math help us understand this situation?*

    + *Do you think that this is fair or unfair? Why?*

    + *What could change this injustice?*

TAKING ACTION

Use the student conversation in response to *What could change this injustice?* and student noticing from the image-based posters around protest marches and so on to introduce ways that people speak up about injustice. Plan with students how you will take action together based on what you have learned. You might take inspiration from these resources:

- "How to bridge that stubborn pay gap," by Claire Cain Miller, *The New York Times*, January 15, 2016 (https://nyti.ms/3Ivvx7u)

- "Actions and Words; They Bring Change" public learning plan from Learning for Justice (https://bit.ly/3IvcoCF)

(Continued)

(Continued)

For example, students could create infographics to communicate their findings to the rest of their school community. These could be posted in school hallways to prompt conversations among students and also between students and their parents/caregivers, particularly around pick-up time when parents might come to the classroom. Or students could make posters including fractions representing the different wages to prompt student agency.

COMMUNICATING WITH STAKEHOLDERS

Before you begin, you may want to email your principal to let them know you will be teaching this lesson. You might also reach out to students' families to provide them with information on what the lesson will cover. Encourage students to share what they are learning with their families. One of the lesson authors, Trisha, heard from students' family members that their children brought up the wage gap during dinner discussions and some families shared pictures of themselves and their children at Women's Marches. You may also invite families to participate in the lesson's classroom experiences.

ONLINE RESOURCES

 Available for download at **resources.corwin/TMSJ-UpperElementary**

Data Suggesting Wage Inequalities

The table shows the average amount that a person from each group is paid compared to the average amount a white man is paid.

	Asian Women	Latina Women	Black Women	Women (Inclusive of all racial and ethnic groups)
1970	50¢	50 ¢	45 ¢	60 ¢
1980	50¢	50 ¢	55 ¢	64 ¢
1990	70¢	50 ¢	60 ¢	72 ¢
2000	69¢	50 ¢	60 ¢	77 ¢
2010	80¢	55 ¢	65 ¢	81 ¢
2020	87¢	58 ¢	63 ¢	82 ¢

Data is compiled from multiple sources and simplified for student use.

AAUW. (2019). Fast Facts: The Gender Pay Gap. https://www.aauw.org/resources/article/fast-facts-pay-gap/

United States Department of Labor, Women's Bureau. (2017). Earnings. https://www.dol.gov/agencies/wb/data/earnings

Worksheet 1: Data Suggesting Wage Inequalities

ABOUT THE AUTHORS

 **Izzy Hendry,** originally from Atlanta, Georgia, has taught in San Francisco for 5 years. She is in the process of becoming a social justice educator. Izzy is inspired by educators who, rather than reproducing an oppressive system, sharpen their students' critical skills by teaching them to critique and disrupt it.

 **Trisha Huynh** is originally from the Bay Area. She now calls San Francisco "home." Trisha went into teaching knowing that she wanted to be a social justice educator. In this endeavor, she regularly incorporates culturally relevant pedagogy through current events and social justice issues into her lessons. She believes that schooling should be focused on students' communities by analyzing them through a critical lens and working to better the world around them.

 **Emma Gargroetzi** is a postdoctoral fellow at University of Texas at Austin. Emma was introduced to racial justice work by her parents who met in the Civil Rights movement, and her work draws on experience as a special educator, mathematics teacher, and mathematics teacher educator in New York, California, and Texas.

- I know when people are treated unfairly, and I can give examples of prejudiced words, pictures, and rules. (Justice 12)

- I know that life is easier for some people and harder for others based on who they are and where they were born. (Justice 14)

- I will work with my friends and family to make our school and community fair for everyone, and we will work hard and cooperate in order to achieve our goals. (Action 20)

MATHEMATICS CONCEPTS

- Reason with fractions, percentages, and/or informal ratios to make comparisons across groups.

- Use the four operations to make sense of real-world data.

- Read and interpret data in tables.

MATHEMATICS PRACTICES

- Make sense of problems and persevere in solving them.

- Construct viable arguments and critique the reasoning of others.

- Model with mathematics.

LESSON 6.3 MODELING LIBRARY FUNDING

Hyunyi Jung and Megan Wickstrom

SOCIAL JUSTICE CONNECTION

The social justice basis for this lesson involves both micro-level and macro-level issues. At the micro-level, we chose to discuss books in school libraries as a topic that is familiar and relevant to students' lives. When students consider the different numbers and types of books in three different schools as well as their own, we hoped that students would notice the disparity of educational resources and contemplate where the disparity comes from. By discussing how to fairly distribute new books to schools, students would brainstorm what actions they can take to make changes to the unequal situations that might have happened in their school, community, and society.

DEEP AND RICH MATHEMATICS

Students will make assumptions about the usefulness of data provided for creating a method. They will mathematize given situations using fractions, ranking or rating, or their newly created mathematical methods. Students will also communicate their problem-solving processes to others and refine developed methods as they solve the problem and share the solution with a larger community.

Resources and Materials

- Access to school library

- Video: "How America's Public Schools Keep Kids in Poverty" by Kandice Sumner, TED (https://bit.ly/3lMxNh4)

- Teacher Resource 1: *Three Images of Library Books*

- Teacher Resource 2: *Data Tables*

LESSON 1 FACILITATION

Introduction

Launch Part I (10 minutes)

- In this part of the lesson, the students are introduced to a problem context and invited to consider how the problem is situated in a broader social justice context.

- Orient students to the issue of library resources: *Today we will discuss how books can be distributed to school libraries.* Introduce the following three pictures to students and ask:

    + *What do you notice?*

    + *What questions do you have?*

Source: unsplash.com/@kazuend

Source: makasana/iStock.com

Source: Yevhen Roshchyn/iStock.com

Explore (20 minutes)

- **Optional:** Have students watch the TED video, "How America's Public Schools Keep Kids in Poverty," by Kandice Sumner (https://bit.ly/3lMxNh4). Discuss the following:

    + *What do you notice?*

+ *What do you wonder?*

+ *Why do you think some schools have more old books than others?*

- Introduce this problem:

Imagine you are selected as representatives on a student library advisory committee. You are in charge of proposing a method or procedure to fairly distribute new books to three different schools. The overarching question we will explore today [or in our next lesson] is: **Which school libraries need books, and why?**

Other students on the library advisory committee have proposed distributing books in the following way:

+ *Amy at Apple Tree: "I think we should distribute an equal number of books to each school."*

+ *Ken at Blue Mountain: "In our school, we need some new science books because of new science projects."*

+ *Hanna at Cool Valley: "I think each school should receive new books depending on the number of students."*

Do you think the three methods suggested are fair? Why or why not? What might you consider when you are distributing the books?

- Have students brainstorm in small groups ideas for what kinds of information they would want to consider when deciding how to distribute the books.

Summarize (5–10 minutes)
- Have each group of students share their ideas with the class.

- Inform students that they will use these ideas in their next lesson to explore this problem further.

LESSON 2 FACILITATION
Modeling

Launch Part 2 (20 minutes)
- Briefly remind students of their discussion from the previous lesson, the premise (library advisory board), and the types of things they said they would consider in deciding how to distribute books. Explain that when people make decisions, they rarely have all the information they might want. For today's activity, they will have the information provided in Figure 1 to base their decisions on. (To note, editable versions of Figures 1 and 2 are provided in Teacher Resource 2: *Data Tables*.)

Figure 1 provides friendly whole numbers to provide to students. However, you may choose to modify these data in a few possible ways depending on your focus. If you prefer messier data, you may adjust each value slightly. If you want to emphasize fractions, we have provided an alternative version of the table in Figure 2. You could also convert to decimals. Alternatively, you may wish to wait and see if students convert to fractions or decimals on their own.

- Consider the data table in Figure 1 and ask the students to share some initial reactions to the following prompts:

    + *What do you notice from the data provided in the table?*

    + *Which library would you want to go to, and why?*

    + *Why do you think each school has a different number of books?*

Figure 1. Data Table That Includes the Number of Books in Three School Libraries

	Apple Tree	Blue Mountain	Cool Valley
Number of fifth-grade students	80	40	40
Total number of fifth-grade books in the library	400	1,600	800
Number of **old** fifth-grade books	240	160	200
Number of **fantasy** fifth-grade books	160	600	160
Number of **science and history** fifth-grade books	100	400	100
Number of **multicultural or anti-bias** fifth-grade books	80	100	40

Figure 2. Data Table That Includes the Number of Books in Three School Libraries Using Fractions

	Apple Tree	Blue Mountain	Cool Valley
Number of fifth-grade students	80	40	40
Total number of fifth-grade books in the library	400	1,600	800
Fraction that are **old**	$\frac{3}{5}$	$\frac{1}{10}$	$\frac{1}{4}$
Fraction that are **fantasy**	$\frac{2}{5}$	$\frac{3}{8}$	$\frac{1}{5}$
Fraction that focus on **science and history**	$\frac{1}{4}$	$\frac{1}{4}$	$\frac{1}{8}$
Fraction with a **multicultural or anti-bias** focus	$\frac{1}{5}$	$\frac{1}{16}$	$\frac{1}{20}$

- After students share ideas, share the problem:

  Imagine that the library advisory committee in a school district has 100 new fifth-grade books to distribute to the three schools in the district. Make a plan to distribute the books and write a proposal to the committee explaining how you have decided the number and types of new books that can be distributed to each school.

  Your proposal should:

  + *List how many of the 100 books each school should get.*

  + *Explain how you used math to reach your recommendation.*

  + *Explain how the committee could use your mathematical approach with other schools.*

- After reading the problem, ask students to briefly share responses to the following two prompts:

  + *What do you know about the problem?*

  + *What are some assumptions we might make to solve the problem?*

TEACHER REFLECTION

When implementing this task, students naturally gravitate to fairness or what seems just and these beliefs often guide the mathematics choices they make. During implementation, it is important to step back and reflect on the bigger questions of *What does fairness mean in this situation?* and *Why does your approach seem fair?* For us, this allowed for students to glimpse one another's perspectives, to realize that we make different choices based on what we value, and to help incorporate multiple perspectives outside of just their own.

Explore (30–50 minutes)

- Let students work in groups to use the information in Figure 1 to develop a method. Visit each group, note their strategies, and pose questions such as these:

  + *How can you use mathematics to understand the different libraries?*

  + *How can you use mathematics to fairly distribute new books to the three libraries?*

  + *What are you going to recommend and why?*

- Provide time for groups to write their proposal to the committee.

- Depending on the progress groups are making, you may want to pause and have groups share rough versions of their thinking thus far.

- If students are new to the modeling process, you may choose to provide additional scaffolding by discussing different ideas and then making class decisions about what approaches to take. However, this will limit the diversity of approaches groups take.

Summarize (20–30 minutes)

- Have each group share their recommendations and their method. Encourage students to ask questions about things the group did and did not consider and why they think their method is fair.

- After all groups have presented, discuss the following questions:

  + *How are the groups' methods similar and different?*

  + *Is it fair for one school to receive more new books than other schools?*

  + *We had certain information about the books in each library. What other information would you want future committees to have when distributing new books?*

  + *What strategies did you find most effective for fairly distributing the books?*

Possible Extension (45+ minutes)

- Work with your school library to develop an inventory of the books in your library. Depending on the information available, you may be able to break the books down into categories similar to what is shown in Figure 1. Alternatively, you may have students work collectively to estimate the number of books in the library. For example, they may count how many are on a shelf and then multiply by the number of shelves.

- If possible, collecting similar data about another school and/or library would allow for a comparison across schools, similar to the previous activity. Discuss what you found, focusing on questions such as these:

  + *What do you notice and what changes do you want to propose to your school?*

  + *What are ways that you can work with others to make schools fair for everyone so that they have equal access to new books?*

TAKING ACTION

As the student library advisory committee, students are in charge of proposing a method or procedure to fairly distribute new books to three different schools. After exploring the data table and proposing a method, students go to their school library, inventory the books, and categorize them to make their school library data table. Students compare their data table with the data provided in this lesson (or with that of another school library) and propose changes they want for their school library.

COMMUNICATING WITH STAKEHOLDERS

Students may propose their methods to the school administrators and other stakeholders. Students can collaboratively work to share the resource disparities they notice and propose fairer ways to distribute new books (and types of books) to schools.

ONLINE RESOURCES

online resources

Available for download at **resources.corwin/TMSJ-UpperElementary**

▲
Teacher Resource 1: Three Images of Library Books

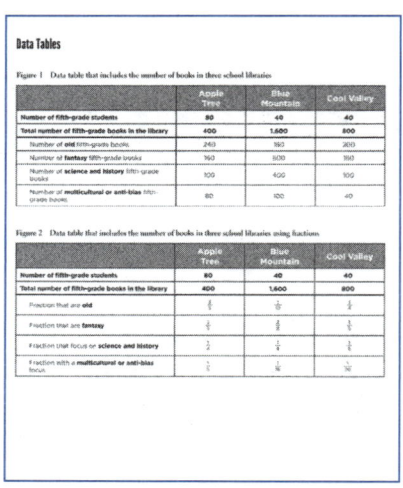

▲
Teacher Resource 2: Data Tables

ABOUT THE AUTHORS

Hyunyi Jung has studied ways to connect mathematics with students' lives through mathematical modeling and culturally sustaining mathematics pedagogy. As a first-generation college student and bilingual learner, she has been interested in the ways in which a broad social system influences mathematical learning and teaching.

Megan Wickstrom is an associate professor of mathematics education in the Department of Mathematical Sciences at Montana State University where she teaches current and future K–12 teachers. As a former middle school mathematics teacher, she finds joy in working alongside classroom teachers to develop tasks that are relevant and important to students and our communities. She works collaboratively to develop tasks that honor students' lived experiences, support their mathematical identities, and allow them to see the world through different perspectives.

LESSON 6.4 THE VALUE OF A SCHOOL LUNCH

Rebecca Ellis, Debasmita Basu, Bethany Chan, and Frances K. Harper

SOCIAL JUSTICE CONNECTION

Research from the U.S. Department of Agriculture shows that 6.5% of American households with children are food insecure. This means that 2.4 million households contain children who are food insecure. This lesson engages students in analyzing the importance that a free lunch can play in a student's life. Through analyzing needs and wants using fractions, students are encouraged to examine the societal and systemic reasons behind food insecurity. In the closing to the lesson, students are asked to evaluate, discuss, create, and possibly participate in solutions addressing the issue of food insecurity and how the problem influences everyone.

DEEP AND RICH MATHEMATICS

Exploring the nutritional value of their school lunches will require students to understand the situation using proportional reasoning and part-to-part and part-to-whole fraction comparisons. They must make decisions about measuring (including making estimates) in order to define a whole and equal parts of that whole (i.e., a fraction). They will rely on understanding that fractions can only be compared when they describe the same whole to analyze and compare the quantity of fruits, vegetables, and dairy (and optionally, grain and protein) in a typical school lunch.

Resources and Materials

- Worksheet 1: *Wants vs. Needs*

- Worksheet 2: *What Fraction of Your Recommended Daily Value Is in Our School Lunch?*

- Teacher Resource 1: *Incorporating Decimals via Money Calculations* (Optional Extension Activity)

- Student Resource 1: *What Is the Cost of School Lunch?* (Extension Worksheet)

- Video: "The Cost of Food Insecurity in Schools," from the Health Forward Foundation (https://bit.ly/3rQcw9U)

- Measuring cups (quarter-cups or half-cups recommended)

- Balance or other weighing tools (optional)

SOCIAL JUSTICE OUTCOMES

- I know that life is easier for some people and harder for others based on who they are and where they were born. (Justice 14)

- I will work with my friends and family to make our school and community fair for everyone, and we will work hard to cooperate in order to achieve our goals. (Action 20)

MATHEMATICS CONCEPTS

- Focus on real-world contexts for understanding fraction concepts conceptually.

MATHEMATICS PRACTICES

- Reason abstractly and quantitatively.

- Use appropriate tools strategically.

- School lunch menu (to help think about which days would be ideal for this unit)

  + Either from your own school or from Figure 2 in this lesson (from School Nutrition Association, 2012). Before copying this sample, we recommend removing the column that labels the food categories.

- School lunch or access to school kitchen

- Menus from local restaurants (We recommend teachers bring these in or ask students to in preparation for this lesson if you plan to do the extension activity.)

LESSON 1 FACILITATION

Exploring School Lunch

Launch (25 minutes)

- Begin by asking your students, *Who's hungry **right now***? There are likely to be a few hands raised.

- Conduct a short discussion about how much time there is from now until lunch/snack time and reflect on what life would be like if students didn't get a lunch break or a chance to eat lunch.

- Distribute page 2 of Worksheet 1 (*Wants vs. Needs*; see Figure 1).

TEACHER NOTE

Throughout this lesson, you can incorporate your own experiences with food insecurity and, if not that, relate the issue of food insecurity in relation to your school or community. Be aware that some of your students may be experiencing food insecurity. Be prepared to hear and validate their experiences while also supporting their needs.

- Conduct a sorting activity in which students place items (sleep, breakfast, games, etc.) into four categories (I need this to survive, This makes my life better, I enjoy this, and I don't need this at all)

  + *Small-group version:* Cut out sets of the 24 item boxes and have students work in groups to sort the items into the four categories.

  + *Whole-class version:* Project the table onto the front board. Students write the items on sticky notes, which they then bring up to the board.

Figure 1. Wants vs. Needs Card Sort

Lunch	Water	Warm clothes
Sleep	Housing	Friends
Dessert	Games	Exercise
Air to breathe	Recess	TV
Breakfast	Cell phone	Umbrella
Snacks	Pillow	Bathing
Dinner	Allowance	

- Regroup the class for a share-out and discuss: *What makes something "necessary" for survival?* Draw special attention to the importance and role of food in their lives.

- Remind students that it is not just any food that is okay. We need to have a nutritious balance of foods to fuel our bodies. Explain to students that the average 10-year-old child needs 1800 calories per day to maintain a healthy weight, grow, and develop. Ask students for their understanding of what the balance of foods might consist of. Have some students share their ideas.

- Continue by building on student responses. Highlight that these calories should come from a variety of sources, including dairy, fruits, and vegetables. For growth and development, it is recommended that children eat 4 half-cups of fruit, 6 half-cups of vegetables, 6 half-cups of dairy, 5 ounces of protein, and 6 ounces of grain every day. However, not all children can access these foods.

TEACHER NOTE

This is an important opportunity to emphasize the meaning of a fraction as a number that relates a part to a whole. In this lesson, the whole is defined by the recommended daily serving, and parts are measured in units of half-cup or an ounce. For Lesson 1, you might address the possible confusion about using a half-cup as the unit of measurement. For example, you could ask, *How many cups of fruit is 4 half-cups?* while emphasizing the relationships between the half-cup and the whole cup of fruit.

For a simpler version of this lesson (i.e., more appropriate for third grade), focus only on the foods that can be measured using half-cups. If you have access to a balance, then you can include the protein and grains (i.e., more appropriate for fourth grade).

Explore (20–25 minutes)

- Place students in small working groups of three to four. Provide each group with copies of a school lunch menu (either from your school or the menu in Figure 2). Ask students to work together in their small groups to:

  + *Categorize the foods into different nutritional categories: fruit, vegetables, dairy, grain, or protein.*

  + *Discuss what counts as what (e.g., Are sweet potato fries vegetables? Are peas vegetables? What is "orange juice"?). Consensus is not required for this discussion.*

Figure 2. Sample School Lunch Menu

Minnesota Sample Cycle Menu

SAMPLE CYCLE MENU GRADES K-5

	MON 1	TUES 1	WED 1	THURS 1	FRI 1
MEAT/MA	Turkey Corndog [W]	Sweet & Sour Chicken [R]	Turkey Sausage	Apple Cider Stew [R]	Turkey Sloppy Joe [R,W]
GRAIN		Brown Rice [W]	French Toast Sticks [W,2 ea]	Dinner Roll [W]	
VEG	Corn on the Cob [L]	Broccoli	Roasted Squash [R,L]	Romaine Salad [1c]	Sweet Potato Fries
VEG	Marinated Black Bean Salad	Carrots	Sliced Cucumbers [L]		Celery Sticks
FRUIT	Watermelon Wedges [L]	Fresh Apple Slices [L]	Orange Juice	Orange Smiles	Juicy Pears
CONDIMENTS	Ketchup, Butter	Ranch, Yogurt Dip	Maple Syrup	Ranch, Butter	Ketchup, Hummus
	MON 2	**TUES 2**	**WED 2**	**THURS 2**	**FRI 2**
MEAT/MA	Cheese Pizza [W]	Hamburger on Bun [W]	Vegetarian Chili [R]	Crunchy Chicken Wrap [R,W]	Fish Sandwich [W]
GRAIN			Cornbread		
VEG	Romaine Salad	Lettuce & Tomato		Potato Wedges	Carrot & Celery Sticks
VEG	Beets 'n' Sweets [R,L]	Fiesta Beans & Rice [R,W]	Fresh Broccoli		Creamy Coleslaw [L]
FRUIT	Juicy Pineapple	Ripe Red Grapes	Applesauce	Orange Smiles	Fresh Kiwi
CONDIMENTS	French dressing	Ketchup, Mayo	Ranch, Butter, Honey	Ketchup	Ketchup, Tartar Sauce
	MON 3	**TUES 3**	**WED 3**	**THURS 3**	**FRI 3**
MEAT/MA	Chicken Sandwich [W]	Chicken Gravy	Meatballs	Chicken Salad on Roll [W]	Cheese Quesadilla [W]
GRAIN		Dinner Roll [W]	Spaghetti & Breadstick [W]		
VEG	Leafy Spinach [1/2 c]	Mashed Potatoes	Tomato Sauce	Carrot & Jicama Sticks	Black Bean Salsa
VEG	Corn Edamame Salad [L]	Spring Salad Mix [R]	Green Beans	Roasted Chickpeas [1/4 c]	Broccoli
FRUIT	Fresh Strawberries	Crazy Mixed-Up Fruit	Golden Peaches	Fresh Pears	Fresh Banana
CONDIMENTS	BBQ Sauce, Ketchup	Butter	Parmesan, Butter	Ranch	Ranch

All serving sizes of fruit are 1/2 cup. Serving sizes of grains are 1-2 oz. Condiments are 1-2 Tbsp.
All serving sizes of vegetables are 1/2 cup (1 cup for leafy greens) unless noted in subscript.
All meals include skim or 1% white milk. Nutrient analysis available on the reverse side.
R = Recipe available at www.health.state.mn.us/schools/greattrays under "Menu Planning"
L = Local food available in many regions of Minnesota
W = Whole grain-rich | Dark Green | Red/Orange | Legumes | Starchy |

GREAT TRAYS

Minnesota Department of
Education

Source: School Nutrition Association (2012). For entire Sample Cycle menu, go to http://www.health.state.mn.us/schools/greattrays/pdfs/SampleCycleMenu.pdf

Summarize (10 minutes)

- Regroup as a whole class. Call on students from each group to share out what food items they included in one of the nutritional categories.

- As students share out, check for agreement or disagreement from the class. Note which food items were more contentious and remind students that some foods can fall into two or more categories.

- *Optional homework assignment*: Reflect on the school menu. Do you get a good balance of foods from your school lunch?

LESSON 2 FACILITATION

Is School Lunch Nutritious?

Launch (10 minutes)

- In small groups, ask students to discuss this guiding question: *Does the school lunch provide a good balance of important foods?*

- As a whole class, review how many nutrients an average 10-year-old child needs for a healthy diet. Write on the board: *The average 10-year-old should eat 4 half-cups of fruit, 6 half-cups of vegetables, 6 half-cups of dairy, 5 ounces of protein, and 6 ounces of grains every day.*

TEACHER NOTE

It is important that students understand that the "whole" changes by the type of food. For example, 2 half-cups of fruit are $\frac{1}{2}$ the daily recommended serving of fruit, but 2 half-cups of vegetables are only $\frac{1}{3}$ the daily recommended serving of vegetables. For third grade, you can emphasize writing equivalent fractions that refer to each of the different wholes (e.g., 2 half-cups of fruit can be written as $\frac{2}{4}$ or $\frac{1}{2}$ of the recommended daily serving; 3 half-cups of vegetables can be written as $\frac{3}{6}$ or $\frac{1}{2}$ of the recommended daily serving).

For fourth grade, you can challenge students to express different addition sentences that show how one could get the daily recommended serving (e.g., $\frac{1}{4} + \frac{1}{2} + \frac{1}{4} = 1$ could be used to show eating a quarter-cup of fruit at breakfast, a half-cup of fruit at lunch, and a quarter-cup of fruit as a snack). Use this discussion to re-emphasize that the "half-cup" is the standard unit of measurement, not the fraction of the recommended daily serving (i.e., the whole). *Important:* All uses of fractions should be connected back to the context (i.e., the part-whole relationship) to emphasize the meaning of fractions and their operations.

Explore (20–30 minutes)

- Divide students into small groups with three to four students in each group. Explain that today your class is going to measure for themselves the nutrients provided in the school lunch.

- Distribute Worksheet 2 (*What Fraction of Your Recommended Daily Value Is in Our School Lunch?*) (**Note:** If your class will be using both half-cups and a balance, use the full handout. If you just want to measure using half-cups, ignore the last page and measure only fruits, vegetables, and dairy.)

TEACHER NOTE

For measuring fruits, vegetables, and dairy, we recommend measuring using half-cups as the unit. You might ask fourth graders to use $\frac{1}{4}$-cup measuring cups in order to provide opportunities for addition of fractions with common denominators.

- Either through taking a field trip to the cafeteria or by bringing in a few samples from today's lunch meal, have student groups measure and record the quantities of the nutrients from the school lunch that day in the handout.

- In order for students to measure these nutrients, they will first need to categorize the meal into nutritional categories. Student groups may come up with different measurements based on what they determine "counts" as part of that nutritional category.

- Students will then complete the rest of the handout. Students should draw or explain the food item, color in the correct fraction of boxes, and then fill in the blank to complete the sentence.

- In their groups, ask students to discuss: *How does the school lunch help students meet their daily requirements?*

Summarize (10 minutes)

- Regroup students for a whole-class discussion. Suggested discussion prompts are as follows:

  + *Which food(s) did we not measure? What nutrients might they be providing?*

  + *Is there anything else in our food that we might want to eat more or less of?* (vitamins, sugars, fiber, etc.)

  + *How much of the daily nutrients are left over for the student to get elsewhere?* Ask students to calculate this using either subtraction or addition with the change unknown (i.e., a + ? = b). This also helps reinforce the idea that claims need to be backed with numbers and facts. Mathematically, this creates opportunities for students to reason about the size of fractions with like and unlike denominators by referencing the whole (i.e., the recommended daily serving). Questions such as *What do students need to eat more of, outside of school?* can only be answered by establishing whether "more" refers to the number of half-cups/ounces (i.e., the parts) or the fraction of the daily recommended serving (i.e., the whole).

LESSON 3 FACILITATION

Food Insecurity

Launch (20 minutes)

- In small groups, have students reflect on the activities of the previous two lessons. As a guiding question, ask: *How much of the daily nutrient requirements are met by the school lunch?* Encourage students to use their handouts and calculations to support their claims.

- Hold a whole-class discussion that focuses on access to important nutrients.

    + *When it's not a school day, what do you do for lunch?*

    + *If you skipped school lunch, what fraction of your total recommended daily [fruit/vegetable/dairy/etc.] would you likely consume today?* (To answer this question, students will likely work with the equation 1 – [lunch amount] = remaining fraction or they will look at which boxes they did not color in on the handout. However, their calculations may differ depending on what they said they do when they skip school lunch.)

    + *Does your typical dinner help you meet the rest of the requirements?* (Sharing out is optional here. Be mindful that not all students come from food-secure homes.)

- Explain that not everyone can access enough food at home and that schools across the nation are looking to provide free breakfast (and lunch) to students to help.

- Decide what fractions of the nutrients should be consumed and when (e.g., *What fraction of your daily recommended values might you get from breakfast? From dinner? From what types of foods?*)

- Depending on time, you may want to have students look up nutritional values from foods they typically (or like to) eat. This can also be assigned as homework.

Explore (20 minutes)

- For this portion of this lesson, bring in the language and concept of *food insecurity*.

- Watch the video "The Cost of Food Insecurity in Schools" from the Health Forward Foundation (https://bit.ly/3rQcw9U).

- Reinforce the fact that over 2 million American households are not food secure. This means that many, many children may go to bed hungry. Or even if the children get enough food, the adults do not.

- To help think about the range of food insecurity, review Figure 3 with your class.

Figure 3. Range of Food Insecurity

Mild food insecurity	Moderate food insecurity	Severe food insecurity
Worrying about ability to obtain food	Compromising quality and variety of food	Reducing quantities, skipping meals	Experiencing hunger

Source: Contains Parliamentary information licensed under the Open Parliament Licence v3.0.

- The theme of food and justice can be an ongoing discussion during the school year. See Teacher Resource 1 for an optional extension activity incorporating decimals via money calculations to explore food insecurity.

Summarize (10 minutes)
- With your class, discuss:

  + *What are your thoughts about making school lunch available every day (regardless of if there is school)?*

TAKING ACTION

Position students as advocates for change. Have students brainstorm how they can use what they learned to improve access to quality food to all children in the community. For example, students might write a letter or make a video (see the Educational Video Center resource at https://evc.org/) to share with the principal, local school board, state representative, or another individual about the importance of school lunch programs helping students and families meet their daily nutritional needs. They could make informational posters to hang up in the community or flyers to pass out around town. In whatever form of advocacy they choose, students should be instructed to include the language of fractions to support their arguments. You can also encourage students to use visualizations to enhance their product. Furthermore, students may also want to advocate for the following:

- Making lunch available to all students every day regardless of if it is a school day or a student is absent.

- Providing more alternative food options depending on dietary restrictions of the students.

- Making lunch cheaper for students.

- Adding or modifying a breakfast program at the school.

- Establishing an alternative meal program.

- Something else based on their personal concerns for student nutrition.

COMMUNICATING WITH STAKEHOLDERS

We recommend that teachers reach out directly to the cafeteria staff and explain that students are doing an activity measuring their school lunches to see how much of their daily recommended nutritional value they can get at school. Students are not measuring the food as a critique of the school lunch program or the quality of the food. Coordinate with kitchen staff to acquire a few extra meals to bring to the class for measuring purposes or coordinate a field trip to the kitchen to measure food there. To further contextualize the unit and to position the cafeteria workers as sources of knowledge, you may want to reach out to the leaders of the nutritional department to see if any of the menu creators would be willing to come to your class to answer student questions about the decisions behind what goes into a school lunch.

Teachers may want to send a letter to parents/caregivers and/or administrators, such as the following:

On [DATES], our class will be measuring food and using fractions to describe and analyze how much of their recommended daily nutritional value they receive from school lunches. This provides students an opportunity to critically reflect on issues of access to important nutrition as well as fairness and justice using mathematics. Food insecurity is an issue for many Americans, and with COVID-19 it has become even more pronounced. This lesson gives students a chance to reflect on the world and to explore their own privileges, using their own calculations as a basis. Students will also have an opportunity to write a supported letter, using their new facts, on possibilities to expand the school lunch program. I invite you to join our lesson and learn with me and the students about how mathematics helps us to reflect critically on this topic. Please let me know if your students have any allergies or other restrictions for working directly with food items so that I can create a safe and meaningful experience for all students.

ONLINE RESOURCES

 Available for download at **resources.corwin/TMSJ-UpperElementary**

Wants vs. Needs

Instruction for Teachers

Cut out sets of these 24 boxes for your students (the blanks are for students to fill in as they wish, or you can decide what else should be included).

In pairs or small groups, students sort the items into four categories presented on the second page. You may want to instead project the table on the board or screen, and have students write their items on sticky notes to place them into the four categories. If school is virtual, we recommend using a program like Jamboard.

Regroup the class for a share-out.

As a class, discuss: *What makes something "necessary" for survival?*

To lead into the next part of the lesson, draw special attention to the importance of and role of food in their lives.

Lunch	Water	Warm clothes
Sleep	Housing	Friends
Dessert	Games	Exercise
Air to breathe	Recess	TV
Breakfast	Cell phone	Umbrella
Snacks	Pillow	Bathing
Dinner	Allowance	

▲
Worksheet 1: Wants vs. Needs

What Fraction of Your Recommended Daily Value Is in Our School Lunch?

Instructions:
1. Draw or describe the food item from the school lunch that provides that nutrient.
2. Fill in the blank to complete the sentence.

Example:
Today I got 3 half-cups of fruit at lunch.

Today's school lunch provided $\frac{3}{4}$ of total daily fruit.

Draw explain your food item:

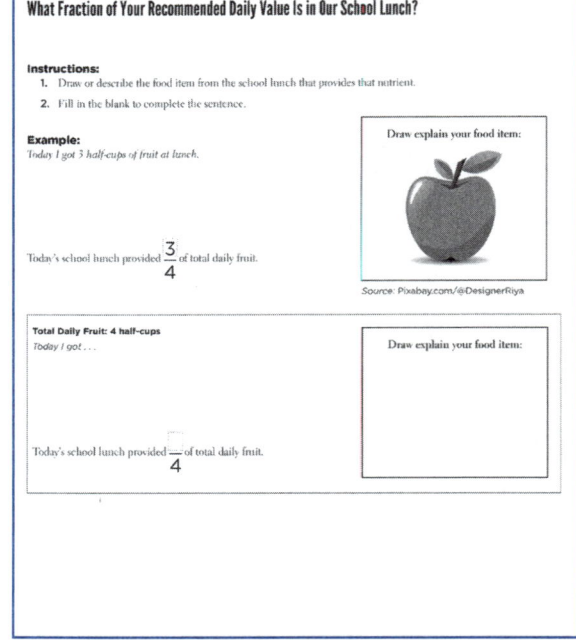

Source: Pixabay.com/@DesignerRiya

Total Daily Fruit: 4 half-cups
Today I got . . .

Draw explain your food item:

Today's school lunch provided $\frac{\_}{4}$ of total daily fruit.

▲
Worksheet 2: What Fraction of Your Recommended Daily Value Is in Our School Lunch?

Optional Extension Activity (25 minutes)

Incorporating Decimals via Money Calculations

This theme of food and justice can be an ongoing discussion during the school year. When you are covering concepts such as

- Multiplication of fractions by a whole number, or
- Using operations with fractions and decimals.

You can revisit food insecurity and incorporate this extension activity.

- Have students either individually write a paragraph or turn-and-talk with a neighbor to discuss what they remember about food and nutrition from the earlier activity.
- Distribute copies of the *What Is the Cost of School Lunch?* Extension Handout from Lesson 6.4 to each student.
- Students work individually or in pairs to answer the following questions:
  + How much does a school lunch cost at your school?
  + If you buy lunch every day (180 days), how much do you spend on lunch?
  + How much would it cost if you only had to pay $\frac{1}{2}$ of the lunch cost?
- Hand out local menus and compare the cost of a restaurant's lunch meal and nutritional value with the school meal.
- Compare the cost of buying in bulk versus buying individually wrapped items to show that a school providing this may be more economical than every family buying individually. This activity also brings in environmental concerns, as there may be less packaging.

▲
Teacher Resource 1: Incorporating Decimals via Money Calculations (Optional Extension Activity)

Extension Worksheet

What Is the Cost of School Lunch?

1. What is the cost of a school lunch?

2. If you buy lunch from school every day (approximately 180 days), how much will you spend on school lunch during a school year?

3. If you paid **half price** for the school lunch, how much would you spend buying lunch every day for a year?

4. Choose a lunch meal from one of the menus.
   a. Which meal did you choose?

   b. How much does the meal cost?

   c. Estimate: How much of your daily recommended nutrients will you get from this meal?

   d. What is the difference in cost between this lunch and the school lunch?

▲
Student Resource 1: What Is the Cost of School Lunch? (Extension Worksheet)

ABOUT THE AUTHORS

Since childhood, **Rebecca Ellis** has been involved with Social Action Tikkun Olam, the Jewish concept of repairing the world, and aims to include social justice in all of her teaching, curriculum design, and assessment development. She recently completed her postdoctoral fellowship with The Connected Biology project, where she researched and developed free and interactive high school evolution education materials. Rebecca was inspired to work on this lesson by her time as an AmeriCorps member at a school where all students received free breakfast and lunch.

Debasmita Basu teaches quantitative reasoning and mathematics at Eugene Lang College of Liberal Arts, The New School, in New York City. Before pursuing her doctoral studies, Debasmita was a high school mathematics teacher in India for 4 years. As a cisgender woman of Color, she aims to design mathematical activities that cultivate students' critical consciousness toward various social and environmental justice issues and help them realize the power and value of mathematics. Previously, Debasmita was involved in the design of two lessons on food deserts and the impact of added sugar on one's health, which motivated her to do this current work.

Bethany Chan is a student in the UOTeach program. She aspires to become a high school mathematics teacher and, eventually, a mathematics curriculum writer. As a first-generation college student and woman of Color, she hopes to inspire more women and people of Color to enter the STEM field through supporting students in realizing the strength of mathematics in lifting up their voices. This particular lesson is connected to her on a personal level, as her mother's job involves providing lunches for a Title I school where many students are provided free breakfast and lunch.

Frances K. Harper taught mathematics and reading across PreK–12 for 9 years in diverse urban settings in Tennessee, Massachusetts, and Kanagawa, Japan. As a white, cis woman and first-generation college graduate, she strives to understand and amplify the perspectives of students and families who have been systematically marginalized in mathematics.

- I will recognize my own responsibility to stand up to exclusion, prejudice, and injustice. (Action 17)

- I know some ways to interfere if someone is being hurtful or unfair and will do my part to show respect even if I disagree with someone's words or behavior. (Action 18)

- I will speak up or do something when I see unfairness, and I will not let others convince me to go along with injustice. (Action 19)

- I will work with my friends and family to make our school and community fair for everyone, and we will work hard and cooperate in order to achieve our goals. (Action 20)

MATHEMATICS CONCEPTS

- Represent and interpret data.

- Reason with number and operation relationships.

- Use real-world contexts for understanding fractions.

MATHEMATICS PRACTICES

- Make sense of problems and persevere in solving them.

- Construct viable arguments and critique the reasoning of others.

- Look for and make use of structure.

LESSON 6.5 MORE THAN AN ATHLETE

Evan M. Taylor

SOCIAL JUSTICE CONNECTION

In his studio album *KOD*, J. Cole pens a song entitled "BRACKETS," which analyzes many of the inconsistencies, shortcomings, and pace of democracy in the United States. One lyric states that "democracy is too slow" and that "it is 2018" so we should be able to vote from our pixelated screens for the things that we want within our country and our specific communities. The song is a critique of current voting practices in the United States, particularly with respect to the negative impact of the process on Black, Indigenous, and other communities of Color. Drawing on the Women's National Basketball Association (WNBA) and National Basketball Association (NBA) All-Star games and the process of player selection, this lesson aims to have students bring their expertise and mathematical knowledge as fans of basketball to analyze voting practices and policies at the local and state levels.

DEEP AND RICH MATHEMATICS

The mathematics of this lesson is focused on data analysis. Students are introduced to fraction concepts and engage in fraction operations as they look at rebounds per game (RPG), assists per game (APG), points per game (PPG), and field goal percentages from beneath and beyond the arc to determine which players of a given list should be chosen and which position they would best play in during the All-Star games.

Resources and Materials

- Worksheet 1: *Roster Selection and Justification Sheet* (1 per group)

- Set of computers, tablets, or iPads

- Website: NBA All-Star Roster (https://www.nba.com/2021-all-star-roster)

- Website: 2021 NBA All-Star Game Guide for background information about the All-Star Game player selection (https://bit.ly/3duYTEO)

- Website: Jr. NBA resource to support understanding of NBA statistics (https://jr.nba.com/how-to-read-a-box-score/)

LESSON FACILITATION

The NBA All-Star Voting Process

Both the NBA and WNBA have All-Star games, and each league allows for fans to vote for which players they would love to see play during the All-Star game. In the past, players have used social media and television commercials to get fans to vote for them, on top of just playing very well on the court and trying to get as many highlights as possible for the fans to see so that they can play in the All-Star game. For about a month, fans (regardless of age) are able to go online and vote for which players they would want to see in the All-Star game. In the past, fans were able to vote for which 12 players they wanted to have on the Western All-Star Team and which players they wanted to play on the Eastern All-Star Team. Presently, players vote for their favorite player and hope that they fall within the top 24 and then those players are chosen by the two players who received the most votes. In this voting process, students (fans) will take on the role of the team captain and will be asked to put together a team of 12 to compare players and see which ones would play best together and have the highest probability of winning the All-Star game together. Through engaging in this lesson, students look at statistics, compare players, and justify their choices with their classmates using mathematical reasoning and statistical analysis. At the same time, they are learning about the voting process.

Launch (30–45 minutes)

- Begin by asking students these questions: *What do you know about voting? When do we vote on things?*

- Have students identify the different opportunities they or their family members get to vote (e.g., presidential election, local city and state election, for their favorite ice cream flavor at the ice cream shop, for their favorite home cooked meal by a parent/caregiver). The goal here is to build from students' existing experiences with voting.

- If basketball and the All-Star Team does not come up, ask the class the follow-up question: *Do any of you follow basketball? What do you know about the All-Star Basketball Team?*

- Have students familiar with the All-Star Team explain it to the class. Here are some resources to share:

  + NBA All-Star Roster (https://www.nba.com/2021-all-star-roster)

  + 2021 NBA All-Star game guide (https://www.nbcsports.com/philadelphia/sixers/nba-all-star-2021-complete-guide-rosters-updated-format-and-more)

Explore: Part 1 (30–45 session)

- Play video highlights of any of the 24 NBA players voted to play in the 2020 NBA All-Star game. In 2020, there were 24 All-Star players. Two players that have intriguing video highlight reels for students of the game are James Harden and Giannis Antetokounmpo:

  + James Harden (https://youtu.be/YeeDMFF9WUk)

  + Giannis Antetokounmpo (https://youtu.be/qdgc5yitWHg)

- Play 2–3 minutes of each video depending on student interest and engagement. Give students time to respond and enjoy the videos and the phenomenal abilities of individual players on the basketball court. Listen to student feedback regarding players who they wish they would have shown highlights of or who they wish were selected for the NBA All-Star game.

- This will serve as a good time to have students help name the actions (field goals, three-pointers, assists, steals, rebounds, turnovers, and blocks) and connect them to the statistics gathered on basketball players. The Jr. NBA website can support student understanding of the NBA Statistics (https://jr.nba.com/how-to-read-a-box-score/), and an NBA Statistical Abbreviation Table is also provided here. Students will have varying levels of basketball knowledge. We've found that there are always a few avid basketball fans in the class. This is the time to tap into their knowledge and love of the game. Encourage these students to lead and facilitate the conversation around naming the actions.

NBA Statistical Abbreviations

Abbreviation	Meaning
PPG	Points per game
APG	Assists per game
RPG	Rebounds per game

Explore: Part 2 (two 45- to 60-minute sessions)

- Review with students the basketball statistical terms discussed the day prior by asking them to share what they know about NBA basketball. During class conversation, ask students to share their knowledge of the NBA All-Star game and the process of a player being chosen to play in the game. In the 2020 NBA season, the NBA decided to continue with the practice of letting the fans vote for who would play in the All-Star game; the two players with the most votes are captains and get to choose teams.

- Hand out Worksheet 1 (*Roster Selection and Justification Sheet*). Students will take on the role of the captains and select a team of 12 All-Star Players, including the captain, from the NBA's 75th Anniversary Team (https://www.nba.com/75). Because a strong team is one that has a range of skills, their team of 12 must have a mix of different positions, including guards, forwards, and a center.

> - **Guards:** There is a point guard and a shooting guard. The point guard is the "leader" of the team on the court. This position requires substantial ball-handling skills and the ability to facilitate the team during a play. The shooting guard is often the best shooter with the capability to shoot accurately from longer distances.
>
> - **Forwards:** There is a small forward and a power forward. The small forward often has an aggressive approach to the basket when handling the ball. They are known to make cuts to the basket in efforts to get open for shots. The power forward and the center make up the frontcourt, often acting as their team's primary rebounders or shot blockers or receiving passes to take inside shots.
>
> - **Center:** The center usually plays close to the basket. They are usually the tallest players on the floor. They are typically skilled at gathering rebounds and setting screens on players.

- Place students in groups of four or five with assigned roles to promote collaborative work. Here are some possible roles and descriptions of the team roles.

> - **Data Collector:** The data collector is responsible for filling out the graphic organizer with the information gathered from NBA.com (https://www.nba.com) on the respective players/positions assigned to the team.
>
> - **Digital Navigator:** The digital navigator is responsible for navigating NBA.com and sharing the information found on the site with the team.

(Continued)

(Continued)

- **Digital Anthropologist:** The digital anthropologist is responsible for making sure that the information gathered is correct. This means that they must be sure that when the navigator is looking for assists per game (APG) that they know that it may be under APG. The anthropologist will want to make sure that this is understood by the data collector as well.

- **Digital Data Analyst:** The digital data analyst will facilitate the conversation around the data that have been collected and recorded.

- **Journalist:** The journalist will share out the data gathered by the team during the selection process.

- Each group member or pair will be given a position (e.g., guards, forwards, and centers) to research. Circulate and guide students to use both mathematical/statistical reasoning and analysis when speaking to their team/community about why a certain player is a better pick than another. The statistics provided often include decimals. Decimals may be a new concept for students. Use the place value chart and place value blocks (with the hundred block representing one whole) to support sensemaking. Next is a sample decimal place value chart from The Curriculum Corner (n.d.).

Decimal Place Value Chart

hundreds	tens	ones	decimal point	tenths	hundredths	thousandths
			.			

Source: The Curriculum Corner (n.d.).

- Encourage students to use the statistics to compare players' performances (e.g., PPG, APG, RPG). *For example, why would you choose to have LeBron James to play point guard? How does your player rank, based on their number of points, rebounds, or assists?*

Summarize: Part 1 (30-minute session)

- Bring the class community back together to share their All-Star Team roster as written on the *Roster Selection and Justification Sheet*. Have student groups rotate to present their argument and justify their rationale for team selection. Encourage the other community members to share if they believe a balanced team is created. This is when students are constructing viable arguments and critiquing the reasoning of others (Standards for Mathematical Practice 3) in reference to the statistical data. Use these sentence starters to prompt active student engagement:

  + *I like how you thought about . . .*

  + *Did you consider . . .*

  + *How did you decide . . .*

  + *What data did you use to . . .*

Summarize: Part 2 (30-minute session)

The purpose of this lesson summary is to get students to think critically about the differences in voting processes and access to candidate information between the NBA All-Star Team and state or national elections.

- Began by asking: *What do you know about the U.S. electoral voting system for state, local, and presidential elections? How is the NBA/WNBA voting system similar to or different from the U.S. electoral voting system?*

- Break up the class into small groups of three or four. Provide poster paper, and ask the following questions to guide students' analysis:

  + *How much information did you collect about each player to determine whether the player would be a good representative or member of your All-Star Team?*

  + *Was it easy to gather information about the players relevant to the roles they may be playing on the team? What information would be helpful to have when making decisions about elected officials? Is the information easily accessible?*

  + *Who was allowed to participate in the NBA All-Star voting in the classroom and in the yearly voting process? Are there restrictions to state and national voting?*

- Have students pause and reflect on whose views were protected and valued in this process and whose views were excluded in this process.

- These questions allow students to begin to compare and develop a critical perspective and analysis of voting and voting rights within their local area. Conduct a gallery review for students to share their findings among groups, and ask these questions:

  + *What wonderings do you have for our own community? What do you want to look up? Who can we ask to gather this information?*

TAKING ACTION

As a possible action step, lead students in writing a letter to a governmental official.

- Introduce students to local government officials and community organizations to whom they can email a letter—those who may impact decisions on voting rights in the area given the data students have collected and the opinions they have expressed about voting inequities and voting suppression in their district or neighboring districts.

- Model how to write a formal letter to an elected official. The letter should include the following:

  + An introduction to the letter writer

  + A clear explanation of the voting right inequity and data to justify its impact

  + An analysis of the data and discussion of why these patterns emerge and its impact

  + At least two ideas the elected official may use to better address the inequities

- Through mock presentations during class, provide support and help students prepare to present their findings at an upcoming council meeting.

- Have student groups present at the upcoming council meeting.

COMMUNICATING WITH STAKEHOLDERS

Encourage other teachers, administrators, and families to join in this activity. Reach out to local community-based organizations on voting rights and voter registration. Contact local museums, city hall, or other entities that have information on the community's voting history. Tapping into community resources will strengthen both the relevance of the unit for students and its impact.

ONLINE RESOURCE

 Available for download at **resources.corwin/TMSJ-UpperElementary**

Roster Selection and Justification Sheet

Position (Circle One)	Guard	Forward	Center
Player Name			
	Points Per Game	Assists Per Game	Rebounds Per Game
1.			
2.			
3.			
4.			
5.			
6.			
7.			
8.			
9.			
10.			
11.			
12.			

Justification: Why do you win the NBA All-Star game?

▲
Worksheet 1: Roster Selection and Justification Sheet

ABOUT THE AUTHOR

Evan M. Taylor is a seventh-year middle school mathematics educator in Indianapolis Public Schools. He is also a mathematics curriculum writer, PhD student at Indiana University–Purdue University Indianapolis, and a SURGE Leadership Academy alumnus. His research is centered on Black males' navigation of various mathematics learning communities, ethnomathematics, and culturally responsive mathematics teaching.

LESSON 6.6 YOUR ACTION SAVES LIVES: COVID-19 AND SYSTEMS THINKING

Jennifer Park

SOCIAL JUSTICE CONNECTION

During the COVID-19 pandemic, communities around the world, even those within the same city, were impacted differently. COVID-19 is not only a health issue but also a social justice issue that heightens decades of "disinvestment and marginalization" (Phillips, 2020) of our most marginalized populations. Health disparities are rooted in inequitable access to basic human needs such as water, healthy food options, a safe dwelling place, stable employment, and a clean living environment. Access to health is often tied to privilege in other areas like class, race, ability, employment, gender identity, and citizenship. While the overarching theme of this lesson takes a systems perspective to health justice, the lesson examines COVID-19 because every student's life has been affected by it to varying degrees. This lesson calls for students to build a critical consciousness of the systems within and interconnected to health care, allowing you and your students to share their knowledge and feelings about the coronavirus and to learn from each other. Student and community empowerment strengthen as they see their capacity to create change while they build a collective understanding of the inequities within health care and take action based on their own needs as well as those of their community.

DEEP AND RICH MATHEMATICS

In this lesson, scholars (students)[1] make sense of quantities and data by considering the units involved and using mathematical tools such as a number line, a hundred chart, base-10 blocks, Unifix cubes, or play money. In discussion, they contextualize the data by connecting them to the systems involved in or interdependent on health care. Scholars also focus on understanding fraction concepts as related to decimals to the hundredths place. As they color the percentages for the racial demographics of their state's population in a decimal chart, they are able to create a visual representation of percentages and to understand that 55% means 0.55 and 55 out of 100. In analyzing how

---

[1] This author refers to their students as *scholars* to show them respect and to name them in positive ways.

the COVID-19 pandemic has impacted people differently, scholars compare two decimals at the hundredths place by reasoning about their size.

Resources and Materials

- Article: "Talking to Children About COVID-19," from *Sesame Street* (https://bit.ly/3DxiwXr)

- Website: The COVID Tracking Project racial demographics for your state (https://bit.ly/3Iu1WLP)

- Highlighters

- Sticky notes

- Student Resource 1: *Decimals Chart*

- Coloring supplies (two to seven different colors, depending on the state)

- Article: "Why COVID-19 Is a Social Justice Issue," by Jai Phillips, *LATogether*, May 15, 2020 (https://bit.ly/3pF20zz)

LESSON 1 FACILITATION

Systems Perspective on Health Care

Launch (20 minutes)

- Begin by introducing systems thinking. Systems thinking is seeing the world as interdependent and interconnected systems, which represent a universal structure of parts and wholes. For example, our solar system is a system; while our solar system is dependent on the sun and connected with other stars in the Milky Way Galaxy, the planets are interconnected systems.

- Have scholars name systems in the classroom, their community, and our world (Figure 1).

Figure 1. Scholar Reflections on Possible Systems

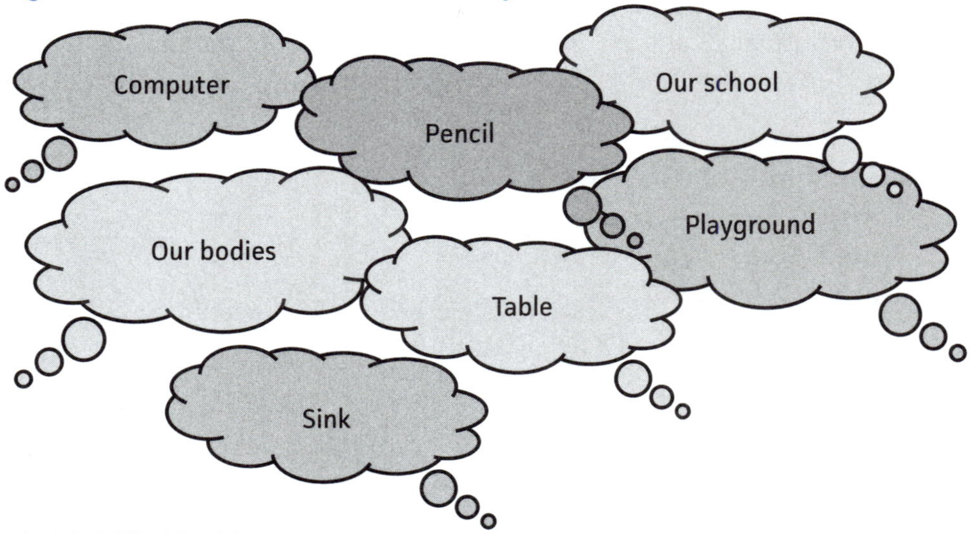

Computer

Pencil

Our school

Our bodies

Playground

Table

Sink

TEACHER NOTE

When scholars aren't certain if something is a system or not, they can leave it up for discussion and keep adding comments on an ongoing discussion board in order to look at lenses from multiple perspectives.

- Now, introduce the topic of this lesson and pose these questions: *What systems make health care? When I say the word "health care," what comes to your mind?*

- As scholars call out the words that come to mind (e.g., "ER", "hospital," "shot"), ask them what systems they are a part of. Some anticipated responses are shown in the example systems thinking map in Figure 2.

- You can also draw scholars' attention to access. Ask: *Who has access to hospitals? What systems impact access to hospitals? To getting shots?* As scholars are sharing, write their ideas down on the systems thinking map so the class can see the parts of each system and how the systems are connected.

- Challenge the whole class to think of as many systems that make up health care together. As you map out the systems, ask questions that allow scholars to see how concepts, people, and places within the system are connected:

  + *How does this connect?*

  + *How is this connected to what is already up on our systems map?*

TEACHER NOTE

The pandemic has affected everyone, but loss, grief, and fear have not affected everyone the same. Throughout the lesson, you'll want to monitor and acknowledge scholars' feelings and provide them space to share, feel seen, and be heard.

Figure 2. Scholars' Concept Map of the Health Care System

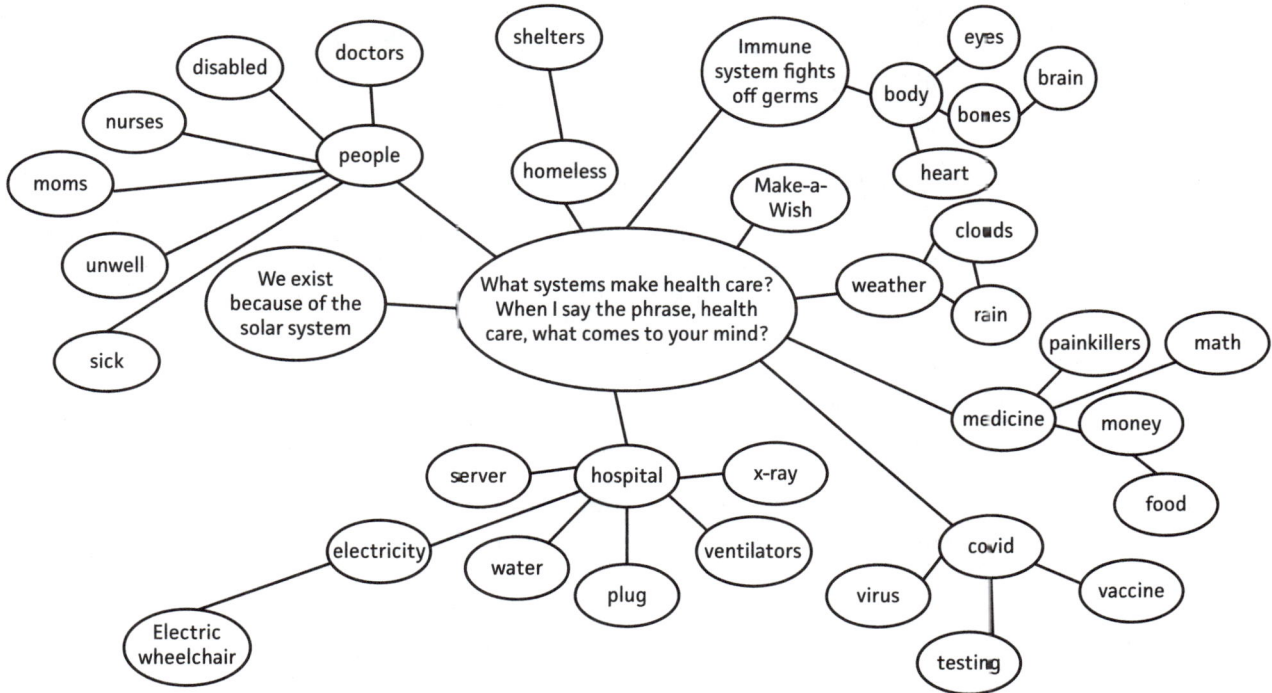

- You may want scholars to sit in one big circle. End the launch by opening up a discussion about what scholars learned from each other while developing the systems thinking map together. Ask: *What do you notice? What do you wonder? How do you feel?* (See Figure 3 for likely responses.)

Figure 3. Scholar Reflections on Systems Thinking

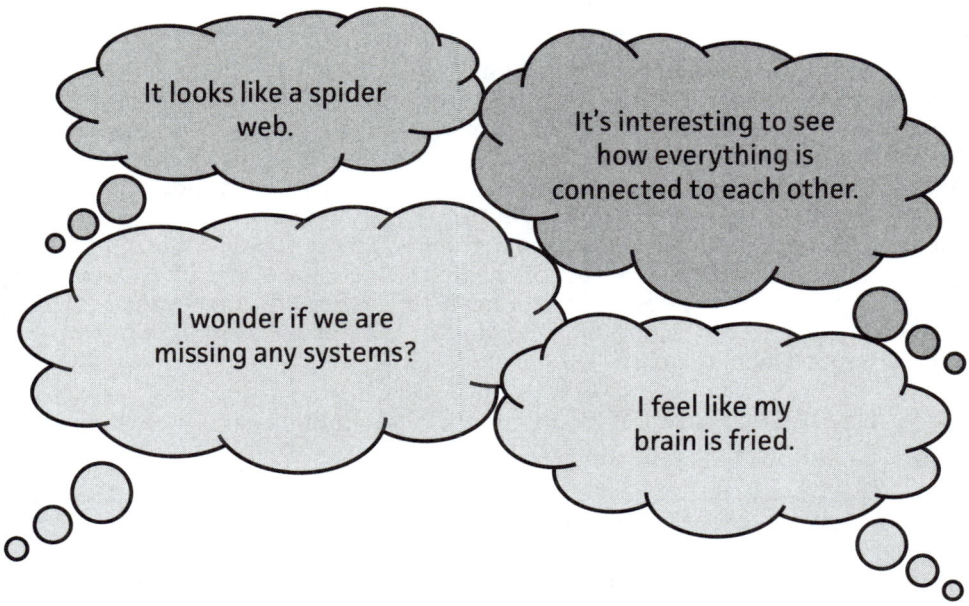

Explore: Part 1 (45 minutes)

Here, scholars will look at health-related data and nonhealth-related data that connect to the systems they named previously. Some examples of data are provided in Figure 4 for your use. You are welcome to add additional data as you learn of or discover them.

- Begin by having scholars form groups of two or three.

- Give each group a slip of paper that shows one of the pieces of data (Figure 4). Give the groups of scholars a moment to discuss.

Figure 4. Data Related to at Least One System on the Health Care Systems Thinking Map

- Children between 6 and 12 years old take 18–30 breaths per minute.[1]

- Angela and JC Johnson, owners of Serving Spoon, a restaurant in Inglewood, California, opened a GoFundMe account and raised $105,151 to keep supporting their employees.[2]

- The Johnsons paid it forward by sponsoring families each week with a $500 check.[3]

- The average cost for a visit to the emergency room was $1,389 in 2017.[4]

- As of February 7, 2021, 17,270,468 people have recovered from COVID-19 in the United States.[5] (**Note:** You can change to up-to-date statistics.)

- As of February 7, 2021, there were 27,532,602 COVID-19 cases in the United States.[6]

- Children's Tylenol Pain + Fever Medicine, Dye-Free Cherry 4 fl. oz. costs $5.47 at Walmart.[7] (**Note:** You can update the price to today's value.)

- In March 2020, 8% of people in the United States were without health insurance coverage.[8]

- In 2011, 214,000 children and teens in Los Angeles, California, had asthma.[9]

Sources: [1] "What is a normal respiratory rate in adults and children?" *Healthline*, March 15, 2018 (https://bit.ly/3DDjcul). [2] "The Serving Spoon (Inglewood) Staff Relief Support," *GoFundMe*, December 8, 2020 (https://www.gofundme.com/f/staff-relief-support/). [3] "What does it mean to be a neighborhood restaurant in a pandemic? How four are surviving," by Jenn Harris, *Los Angeles Times*, December 9, 2020 (https://lat.ms/3ycEv4H). [4] "'Really astonishing': Average cost of hospital ER visit surges 176% in a decade, report says," by Ken Alltucker, *USA Today*, June 24, 2019 (https://bit.ly/3pD7Otr). [5] Worldometer COVID Cases US (https://bit.ly/31HUFHa). [6] "Covid-19 Cases in the United States today – Search," Bing Search, February 7, 2021 (https://binged.it/3iTGRi4). [7] Cost of children's Tylenol at Walmart in December 2021 (https://bit.ly/3EDNRJ7). [8] "Percentage of people without health insurance coverage by selected characteristics: 2019," from U.S. Census Bureau (https://bit.ly/3JYeSK9). [9] Asthma Coalition of Los Angeles County (http://www.asthmacoalitionla.org/).

- Have scholars in each group discuss these questions and write notes on a T-chart. See the following example.

What do you think this means?	What questions do you have?

- Support scholars' sensemaking of the data, encouraging them to represent the data display using mathematical tools of their choice. The mathematical tools they can use are a number line, a hundred chart, place value charts, base-10 blocks, Unifix cubes, or play money.

- The data shown give scholars an opportunity to make sense of multidigit numbers in context. As the statistics are read, ask the following questions:

  + *How did you know how to read this number?*

  + *How much is that?*

  + *How is that value significant? Why?*

  + *Why does this data matter?*

  + *How does this data impact our understanding of health care justice?*

Explore: Part 2 (45 minutes)

- Have each group talk briefly about the following question, *Where does your group's piece of data belong on the systems thinking map?* Then have each group read their piece of data to the class and post it (with a sticky note) on the systems thinking map where they think it belongs.

- As each group presents their data and sticky notes on the systems thinking map, ask them, *Why do you think it belongs there on the map?*

- Encourage other scholars to ask questions or give respectful comments to the presenters. Offer some helpful sentence frames such as these:

  + *I agree/disagree because___.*

  + *I'm not sure what you mean by__.*

  + *I think it could also go ____.*

Summarize (20 minutes)

- Tell scholars that we as a classroom community are growing in seeing our world through a systems thinking perspective. Lead a group discussion with the following prompts:

  + *What does all of this information tell you about health care?*

  + *How do these pieces of data connect to you, to your family, or to your community?*

  + *Is there someone in your family or in our community that knows more about health care?*

- Some possible scholar reflections are shown in Figure 5.

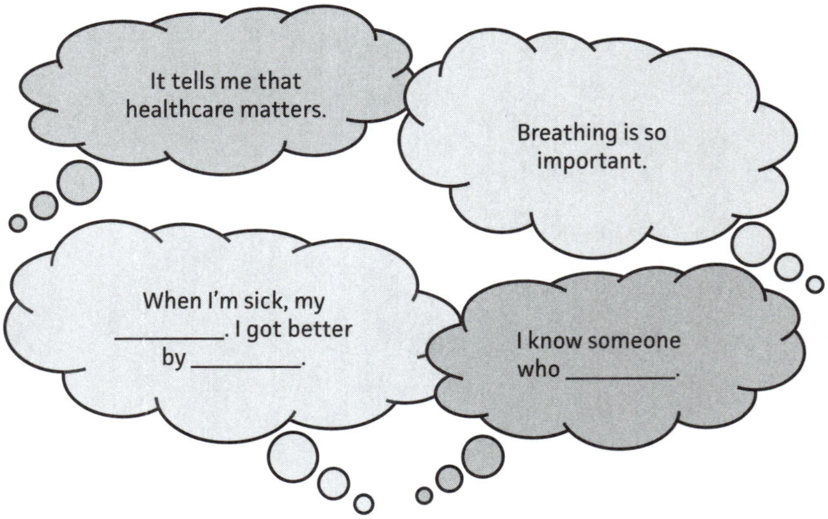

Figure 5. Scholar Reflections on Health Care

It tells me that healthcare matters.

Breathing is so important.

When I'm sick, my _____. I got better by _____.

I know someone who _____.

LESSON 2 FACILITATION

COVID-19

Launch (20 minutes)

- COVID-19 caused rapidly changing situations, loss of daily routine, isolation, and uncertainty, as well as a spread of information overload, rumors, and misinformation. Talking to scholars about what has happened, validating their feelings, and providing them with accurate facts is important.

- Ask scholars questions about what they know about the coronavirus/ COVID and how they feel:

  + *What have you heard or what do you know about the coronavirus?*

  + *What questions do you have? How do you feel?*

- Reassure their feelings and concerns and share your own feelings.

- Explain that the coronavirus is a virus (germ) that can make a body sick. People who have COVID-19 often have a cough, fever, and trouble taking deep breaths, but some people, especially children, can have the virus and may not feel sick at all or have mild or no symptoms. They may know someone who has gotten sick from the virus or had the virus.

- Talk about what is being done. Experts continue working hard to learn more about the virus and the ways it changes to keep us safe. Scientists have created vaccines, which teach your body's immune system to recognize and defend against harmful germs.

- Have scholars share ways to protect themselves and others. Ask: *What are some ways you have protected others and yourself? What are some things you know people have done to get better when they got sick?*

This portion of this lesson was guided by the "Talking to Children About COVID-19" resource from *Sesame Street* (sesamestreetin communities.org/ activities/talking-about-covid-19-with-children/).

Explore: Part 1 (45 minutes)

- Begin the activity by displaying the percentages of racial demographics of your state population, which can be found using The COVID Tracking Project Racial Data Dashboard (https://bit.ly/3Iu1WLP).

- Tell scholars that they are first going to focus on the percentages of the population (the first main column of data).

- Provide scholar pairs with the following hundredths chart. Have them make sense of the percentages for each racial demographic in your state with this prompt: *Let's make sense of each of the percentages. Percent means out of a hundred. On a hundredths chart, each little unit is one percent.*

- Before you have scholars begin, ask them (1) how they might read the decimal numbers and (2) how they might show the population percentages on the decimal chart. Have scholar volunteers color in and label the percentage with the racial demographic. If needed, model how to read the percentage, fill in the hundredths chart, and label it.

0.01	0.02	0.03	0.04	0.05	0.06	0.07	0.08	0.09	0.1
0.11	0.12	0.13	0.14	0.15	0.16	0.17	0.18	0.19	0.2
0.21	0.22	0.23	0.24	0.25	0.26	0.27	0.28	0.29	0.3
0.31	0.32	0.33	0.34	0.35	0.36	0.37	0.38	0.39	0.4
0.41	0.42	0.43	0.44	0.45	0.46	0.47	0.48	0.49	0.5
0.51	0.52	0.53	0.54	0.55	0.56	0.57	0.58	0.59	0.6
0.61	0.62	0.63	0.64	0.65	0.66	0.67	0.68	0.69	0.7
0.71	0.72	0.73	0.74	0.75	0.76	0.77	0.78	0.79	0.8
0.81	0.82	0.83	0.84	0.85	0.86	0.87	0.88	0.89	0.9
0.91	0.92	0.93	0.94	0.95	0.96	0.97	0.98	0.99	1

- Using the hundredths chart that scholars colored in, labeled, or placed blocks on as resources, ask the following questions:

  + *Which racial demographic population has the highest/lowest percentage?*

  + *What else does this data tell you?*

  + *What questions do you have about this data?*

Explore: Part 2 (45 minutes)

- Pose this question to have scholars connect the chart to the systems thinking map: *Where on the systems thinking map might this information about our state's population belong?* The following are possible scholar responses:

  + *Everyone can get a cold or have a runny nose.*

  + *Everyone is human and needs to breathe.*

+ *Not everyone has the same amount of money.*

+ *Everyone was impacted by the pandemic.*

- As scholars name the systems, highlight them on the map and ask follow-up questions:

    + *Does everyone get a cold the same? At the same time and for the same number of days?*

    + *Why do you think some people may have an easier or harder time breathing?*

    + *Why does everyone not have the same amount of money? Does or did everyone get COVID-19 with the same symptoms?*

- Tell scholars that they are going to compare their decimal chart with another that shows the racial demographics of COVID-19 cases to reflect on how the pandemic has impacted people differently. Using racial demographics for your state provided by The COVID Tracking Project (https://bit.ly/3Iu1WLP), display the racial demographics of the COVID-19 cases of their home state on a decimal chart and have scholars compare the charts. Ask the following questions: *What do you notice? What do you wonder?*

- Ask scholars to think of the systems of care that weren't provided for (i.e., the racial demographic that represented the highest percentage of COVID-19 cases). As they think of them, have them count them by holding their fingers close to their chest.

Summarize (20 minutes)

Read aloud and present the following modified quote from Jai Phillips, a program officer in youth development in Los Angeles, along with a picture of a rose that grew from concrete.

> *People are like gardens: they need the right seeds to be planted, watered and nurtured in order to grow, but if they are left alone, they will never produce the full harvest of their potential. For decades, Black people have been starved of the investments and the systems of care and quality environments that people need to grow and to bloom. For even a rose can bloom from the concrete, if we give it the resources, care and love it deserves.*
>
> —Jai Phillips (2020)

Source: dewed/iStock.com

- Have scholars do a think–pair–share activity to answer these questions: *What does a garden need to grow? How are we and people in our community like gardens?*

- Possible reflections are shown in Figure 6.

Figure 6. Scholar Reflections on Gardens and Growing

TAKING ACTION

Tell scholars that an important part of learning is to share what they have learned with others and for them to share their knowledge with you. We create change when we learn from each other and take action that benefits our collective community. Ask scholars what they learned from the data, from the systems thinking map, and from each other and what resources their community needs. Also ask them whom they would want to share their knowledge with, in order to take steps toward positive change for our health care system or to argue for more equitable resources to address health disparities based on community needs. Present their systems thinking map to a stakeholder/decision maker in the health care community. For example, scholars might consider approaching stakeholders or decision makers such as a district school board member, the mayor, a city council member, their school parent–teacher association members, or their principal.

COMMUNICATING WITH STAKEHOLDERS

Teachers may communicate the social justice and mathematics learning goals of this lesson to parents/caregivers by holding a brief informational meeting online or in person and sending a follow-up email. This will provide some time for parents/caregivers to reflect on and think about whether they are willing to allow their children to participate in this unit; also, they may have resources, suggestions, or other information that they would like to provide as input. Teachers may also communicate with their administrators in advance by setting up a meeting to inform them of the unit and to ask for suggestions.

ONLINE RESOURCE

 Available for download at **resources.corwin/TMSJ-UpperElementary**

Decimals Chart

0.01	0.02	0.03	0.04	0.05	0.06	0.07	0.08	0.09	0.1
0.11	0.12	0.13	0.14	0.15	0.16	0.17	0.18	0.19	0.2
0.21	0.22	0.23	0.24	0.25	0.26	0.27	0.28	0.29	0.3
0.31	0.32	0.33	0.34	0.35	0.36	0.37	0.38	0.39	0.4
0.41	0.42	0.43	0.44	0.45	0.46	0.47	0.48	0.49	0.5
0.51	0.52	0.53	0.54	0.55	0.56	0.57	0.58	0.59	0.6
0.61	0.62	0.63	0.64	0.65	0.66	0.67	0.68	0.69	0.7
0.71	0.72	0.73	0.74	0.75	0.76	0.77	0.78	0.79	0.8
0.81	0.82	0.83	0.84	0.85	0.86	0.87	0.88	0.89	0.9
0.91	0.92	0.93	0.94	0.95	0.96	0.97	0.98	0.99	1

Student Resource 1: Decimals Chart

ABOUT THE AUTHOR

Jennifer Park was born and raised in Southern California, and she loves mathematics. As a teacher, she is learning about ways to make mathematics personal for scholars and to create a space where they listen to, care for, and support each other. To Jennifer, becoming a social justice educator entails approaching teaching as a life-giving, humbling act in which she and her scholars challenge the banking method of education.

MATHEMATICS LESSONS TO UNDERSTAND OUR WORLD

CHAPTER
7

We live in a global society. Students in our schools share regional and global responsibility for the planet and must be able to take a global perspective that challenges injustice and destruction wherever they are found. Teaching mathematics for social justice engages students in critically understanding their world and also in acting toward a more peaceful, just, and healthy world.

LESSONS

Lesson No.	Lesson Title	Mathematics Focus Areas	Social Justice Topic	Authors
7.1	*Water Is Our Right, Water Is Our Responsibility*	• Whole Number Concepts and Operations	Environmental Justice	Carolee Hurtado and Iris Franco
7.2	*Upper Elementary Mathematics to Explore People Represented in Our World and Community*	• Whole Number Concepts and Operations • Data Concepts and Statistical Thinking	Cultural and Global Diversity	Lynette Guzmán, Jeff Craig, Eva Thanheiser, Mary Candace Raygoza, and Courtney Koestler
7.3	*Single-Use Plastics*	• Fraction Concepts and Operations (Decimals) • Whole Number Concepts and Operations	Environmental Justice	Carolee Hurtado

- I will respectfully express curiosity about the history and lived experiences of others and will exchange ideas and beliefs in an open-minded way. (Diversity 8)

- I will identify figures, groups, events, and a variety of strategies and philosophies relevant to the history of social justice around the world. (Justice 15)

- I will recognize my own responsibility to stand up to exclusion, prejudice, and injustice. (Action 17)

MATHEMATICS CONCEPTS

- Develop deeper understanding of place-value concepts that support meaningful work with operations.

- Represent and solve problems with multiplication.

MATHEMATICS PRACTICES

- Make sense of problems and persevere in solving them.

- Reason abstractly and quantitatively.

LESSON 7.1 WATER IS OUR RIGHT, WATER IS OUR RESPONSIBILITY

Carolee Hurtado and Iris Franco

SOCIAL JUSTICE CONNECTION

This lesson focuses on environmental justice related to water and in learning about Indigenous-led movements for social and environmental justice through the children's book, *We Are Water Protectors* (Lindstrom, 2020). Students think critically about the meaning of environmental justice in terms of what it means not only to speak up for their individual communities but also to empower the voices of overlooked people and animal friends. In this lesson, students consider steps they can take to reduce their own water usage and they express their opinions through writing to others outside of class.

DEEP AND RICH MATHEMATICS

Students use mathematics to support them in formulating their own opinions and to draw their own conclusions. In this lesson, students add and multiply large numbers by using data from existing sources and estimating their own water usage. The mathematics concepts for Lesson 1 include modeling with mathematics, estimation, writing and interpreting numerical expressions, and calculating averages. The mathematics concepts for Lesson 2 include making sense of problems and persevering in solving them, constructing viable arguments and critiquing the reasoning of others, and using place-value understanding and properties of operations to perform multidigit arithmetic.

Resources and Materials

- Book: *We Are Water Protectors* by Carole Lindstrom

- Video: "Why Care About Water?" from *National Geographic* (https://bit.ly/3ya2OjG)

- Image from article: "Judge orders Dakota Access Pipeline spill response plan, with tribe's input," by Phil McKenna, *Inside Climate News*, December 5, 2017 (https://bit.ly/3oEjXyN)

- Video: "Dakota Access Pipeline Explained: What Is It, Why Are People Protesting It, and What Happens Next?"

by Mythili Sampathkumar, *Independent*, January 24, 2017 (https://bit.ly/3ybZdBH)

- Maps from article: "These maps help fill the gaps on the Dakota Access Pipeline," by Lyndsey Gilpin, *High Country News*, November 5, 2016 (https://bit.ly/3IBPwkY)

- Article: "Oil, water, and steel: The Dakota Access Pipeline," *Earth Justice* (https://bit.ly/31MX1Vv)

- Article: "Keystone Pipeline History," Mark Hefflinger, *Bold Nebraska*, November 7, 2019 (https://bit.ly/3dF4isO)

- Worksheet 1: *Say, Mean, Matter Graphic Organizer*

- Teacher Resource 1: *How Much Water Do We Use?* (PowerPoint slides)

- Sticky notes (3 colors, 1 of each color per student)

- 1-gallon container (e.g., milk container)

- Teaspoon

- 5-gallon water container (optional)

LESSON 1 FACILITATION

How Much Water Do We Use?

Launch (20 minutes)

- Following the PowerPoint slides in Teacher Resource 1 (*How Much Water Do We Use?*), begin the lesson by posing the following questions (slide 2):

  + *¿Cuáles actividades requieren aqua? (What activities require water?)*

  + *¿Cuánta agua cree que usa en un día? (How much water do you think you can use in 1 day?)*

- Using a think–pair–share instructional routine, ask students to "think" about daily activities requiring water and how much water they use in 1 day. This "think" time should be done by students individually with no talking. After a couple of minutes, "pair" students to "share" with each other how much water they think they use in 1 day.

- After students have shared with their partners for a few minutes, ask each student to write their estimate on a sticky note.

- Display the sticky notes on a wall or whiteboard for all students to see.

- Engage in a whole-group discussion by asking students to share and explain how they arrived at their estimated daily water usage.

- After students share during the whole-group discussion, present data on the average water usage based on daily activities (slides 3 and 4), which is informed by data from the U.S. Geological Survey (https://water.usgs.gov/edu/activity-percapita.html). Use 1-gallon and 5-gallon containers as visuals to help students make sense of the amount of water used in these activities.

Explore (30 minutes)

- Start by sharing data with students on average water usage based on every-day activities (slides 3 and 4). Return to the think–pair–share instructional routine. Invite students to return to their original estimates considering the data provided and ask them if they would like to revise their estimate. Individual students "think" quietly for a couple of minutes about daily activities requiring water and how they might revise their original estimate. "Pair" students to "share" their revision and strategies used to revise their daily water usage.

- After students have shared with their partners for a few minutes, ask each student to write their current estimate on a different-colored sticky note.

- Display the sticky notes on a wall or whiteboard for all students to see.

- Engage in a whole-group discussion by asking students to share and explain how they arrived at their revised estimate.

- Here you have three options for your next instructional step:

    + Take the average of all student estimates to find the class average,

    + Share that, on average, each person uses 80–100 gallons of water per day (the first statement on slide 5), or

    + Ask students to use their estimate.

- Based on data decided upon from the option you selected, ask students to determine how many gallons of water they use in 1 week. Then have them determine how many gallons of water they use in 1 month. We suggest having students sit in small groups of three to four; although they will be working independently, they can talk to each other, ask questions, and share ideas.

- Engage in whole-class discussion on weekly and monthly water usage. Highlight the different strategies students used and how those strategies are similar and different, mathematically.

- Regroup students into pairs. Extend the question, now asking students to determine the water usage for their family for 1 week (the second question on slide 5).

- Students first determine the total number of people living in their family. Be sure to discuss how a family's water needs may overlap with individual water needs, such as when cooking, gardening, and doing laundry, and to consider how to account for this when estimating water usage.

- Within the pair, have each student share with their partner their strategies for finding weekly water usage for their family. After discussion, each student should put their weekly family water usage on a different-colored sticky note.

- Have student pairs join another student pair to share how they determined their family water usage for 1 week. You should circulate the room, taking note of student strategies.

- In a whole-class discussion, invite a few students or groups to share their strategies for determining weekly family water usage. Highlight the different strategies students used and how those strategies relate mathematically.

Summary (10 minutes)

- Following the whole-class sharing of mathematical strategies to determine their family's weekly water usage, ask students to discuss the following questions (slide 6).

  + *¿Cuáles son algunas maneras de reducir la cantidad de agua que usa?* (*What are some ways you can reduce how much water you use?*)

- Discuss in small groups actions we can take.

- Take Action: Make a Pledge! Place your pledge on chart paper or butcher paper on the wall.

LESSON 2 FACILITATION

Dakota Access Pipeline

Launch Part 1 (15 minutes)

- We are continuing our exploration of the importance of water in our lives.

- Start by watching the video "Why Care About Water?" together (https://bit.ly/3ya2OjG).

- Ask students: *What did you notice? What do you wonder?*

- Orient students to information shared in the video:

  + The amount of usable water is less than 1 teaspoon per gallon of the world's water.

  + Use a 1-gallon container and a teaspoon to provide a visual for students.

Launch Part 2 (30 minutes)

- Return to the book *We Are Water Protectors*. Re-read as necessary with the class. Use the following questions to guide whole-class discussion of the book:

  + *What is the context that the author draws from to write this book?*

  + *What does the black snake represent?*

- Show an image from the protests from the *Inside Climate News* article, "Judge orders Dakota Access Pipeline spill response plan, with tribe's input" (https://bit.ly/3oEjXyN).

- Ask: *What do you notice? What do you wonder?*

- Then ask: *What is the Dakota Access Pipeline?* (Depending on your geographic location, your students may or may not be familiar with the Dakota Access Pipeline. Asking this question helps you ascertain what prior knowledge they may have.)

- Together, watch a short video from the *Independent*, "Dakota Access Pipeline Explained: What Is It, Why Are People Protesting It and What Happens Next?" The video describes the Dakota Access Pipeline and introduces two different perspectives (https://bit.ly/3ybZdBH).

- Use the following questions to guide the whole-class discussion:

  + *Should the Dakota Access Pipeline be constructed?*

  + *What considerations should be taken into account in making this decision?*

- Share with students that we will use data to formulate opinions about the Dakota Access Pipeline and debate.

Explore (90 minutes)

- Display a map of the Missouri River Basin Contemporary Tribal Land and the Proposed Oil Pipeline (see https://bit.ly/3IBPwkY).

- Ask students: *What do you notice? What do you wonder?* Record student responses.

- Share with students that according to the Earth Justice (n.d.) article titled "Oil, water, and steel: The Dakota Access Pipeline," the Standing Rock Sioux Tribe's 8,000 tribal members and 17,000,000 people rely on the Missouri River for drinking water (https://bit.ly/31MX1Vv).

- Continue the whole-class discussion by asking:

  + *Who is impacted by the construction of the Dakota Access Pipeline?*

  + *Will tribal lands be impacted by the construction?*

  + *Can oil pipelines leak?*

  + *What data can we use to determine if the proposed pipeline is safe?*

- Here you have two options to decide the next instructional step:

  + Share with students that oil leaks are an environmental contaminant, potentially coating everything oil touches. It can destroy ecosystems, pollute water, and harm or kill plants and animals.

  + Have students conduct research to learn about the implications of oil spills on land and in water. Add 60 minutes to the lesson if you would like students to conduct their own research.

- Next have students access the Earth Justice (n.d.) article (https://bit. ly/31MX1Vv) to analyze a list of oil leaks that have occurred in an oil pipeline for a 10-year period. An additional resource is the Keystone Pipeline History from Bold Nebraska (https://bit.ly/3dF4isO).

- Distribute Worksheet 1 (*Say, Mean, Matter* graphic organizer) to each student (see Figure 1). Students record what the data "says," "means," and how it "matters" for the Dakota Access Pipeline.

Figure 1. Say, Mean, Matter Graphic Organizer

SAY	MEAN	MATTER
This is a summary of what the text says and/or what the data shows. What happened? What calculations did I make?	This is an interpretation of what the text says, what data shows, and/or the calculations I make. How do I interpret this?	What are the implications? Why is this important? Why does it matter to me or others?

- Return to the original questions: *Should the Dakota Access Pipeline be constructed? What considerations should be taken into account in making this decision?* Have students share their opinion using mathematics as a way to provide evidence for their claims.

- **Activity Option 1:** Students share opinions in verbal and written forms.

- **Activity Option 2:** Students work in groups of four. Have them create a poster to share their group's perspectives on this topic. They should use mathematics, words, and art to express their views. Follow this with a gallery review.

- In the gallery review, two students stay at the poster and engage other students in discussion around their poster while the other two students visit other posters. This is then repeated for the other pairs of students.

In an effort to use more inclusive language, we use the term *gallery review* for what is commonly referred to as a "gallery walk."

Summarize (60 minutes)
- As a class, re-read the book *We Are Water Protectors*. Engage in a whole-class discussion guided by these questions:

  + *What is your pledge or commitment to action?*

  + *What are some ways that we can be water protectors?*

- Provide students opportunities to express their opinions beyond the class discussion.

TAKING ACTION

In addition to the pledge students might make about their personal water use suggested in Lesson 1 (How Much Water Do We Use?), students might extend their pledge to take action in the following ways:

- Collect your own data! Note each time you use water for 1 week. Calculate your water usage. How did this compare to your estimations? (30–60 minutes for strategy sharing and reflection discussion)

- Collect your own data! Note your family's water usage for 3 days. Share your pledge with your family. After 1 week of implementing your pledge, note your family's water usage for 3 days. How does your water usage compare between the first and second data collection? Devote class time for a follow-up lesson/discussion for students to share their results and analyses. Ask students what impact this has had on their family and their class's community collectivism. (60 minutes)

- Put your ideas for reducing water into action! For the next week, take note of every time you reduced your water usage. How much water did you save this week? (30–60 minutes for strategy sharing and reflection discussion)

At the end of Lesson 2 (Dakota Access Pipeline), students are asked again to consider action they might take to be water protectors. Examples of actions students might take include, but are not limited to, writing letters to government officials, writing letters to the Standing Rock Sioux Tribe, or sharing knowledge with family and friends (verbally through informal conversation or through posters, flyers, or a more formal presentation) to help others understand the issue.

COMMUNICATING WITH STAKEHOLDERS

You may want to send a letter to parents/caregivers and/or administrators, such as this one:

Mathematics is everywhere! We will be reading a children's book and connecting mathematics lessons to the story. The book, We Are Water Protectors, *is a story inspired by Indigenous-led movements to protect water from harm. We will consider our own water usage and use mathematics to better understand the perspectives shared in the book and formulate our own opinions of recent events.*

ONLINE RESOURCES

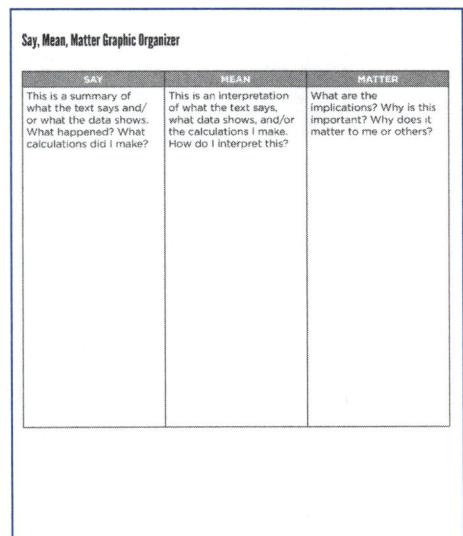

◀ *Worksheet 1: Say, Mean, Matter Graphic Organizer*

▶ *Teacher Resource 1: How Much Water Do We Use? (PowerPoint slides)*

Water
Agua

How Much Do We Use?
¿Cuánto usamos?

ABOUT THE AUTHORS

Carolee Hurtado is a mathematics educator at California State University Channel Islands who is interested in social justice, critical pedagogy, and centering student voice in K–16 classrooms. One of her favorite places to be is on the classroom rug listening to young mathematicians share their thinking. She is excited to collaborate on this project with her colleague Iris Franco.

Iris Franco is a first-generation college graduate, motivated to create classroom communities that empower students to be actively engaged in their learning and the community around them. She is excited to show her students that they are all capable of greatness, and that their backgrounds and experiences contribute a wealth of knowledge. She is delighted to be working with Carolee Hurtado in enriching students' mathematical experiences.

- I have accurate, respectful words to describe how I am similar to and different from people who share my identities and those who have other identities. (Diversity 7)

- I want to know more about other people's lives and experiences, and I know how to ask questions respectfully and listen carefully and non-judgmentally. (Diversity 8)

- I know that life is easier for some people and harder for others based on who they are and where they were born. (Justice 14)

MATHEMATICAL CONCEPTS

- Make sense of large numbers.

- Represent and interpret data.

MATHEMATICS PRACTICES

- Make sense of problems and persevere in solving them.

- Construct viable arguments and critique the reasoning of others.

- Model with mathematics.

- Attend to precision.

LESSON 7.2 UPPER ELEMENTARY MATHEMATICS TO EXPLORE PEOPLE REPRESENTED IN OUR WORLD AND COMMUNITY

Lynette Guzmán, Jeff Craig, Eva Thanheiser, Mary Candace Raygoza, and Courtney Koestler

SOCIAL JUSTICE CONNECTION

By shrinking our world into a village of 100 people and learning about different aspects of their lives—"nationalities," languages spoken, religions practiced (or not!), access to drinking water, living in poverty, etc. (Smith, 2011)—students can explore a global context and engage in critical inquiry about all of humanity's access to basic human needs and distribution of resources. Engaging with these data can affirm students' identities and/or challenge assumptions about who lives in our world and the many facets of their lives, thus empowering everyone to be more aware of and connected to the global community. We center three overarching themes connected to social justice. First is *community*: as teachers and students, we are all members of a classroom community. We are also members of the communities in which we live and the global community of all people. This lesson invites students to build a beloved classroom community, respecting the diversity within the class. It also invites students to reflect on who makes up our global community and how this is similar to and different from the whole that makes up our local community. Second is *critical literacy*: a way of seeing ourselves and our world. It is an approach we take with students in this lesson to ask questions to uncover and unpack assumptions connected to dominant narratives about our world. Third is *representation*: in this lesson, students have opportunities to see how the representations one chooses change the way the data are communicated. In other words, representations are not neutral, just as all mathematics is not neutral.

DEEP AND RICH MATHEMATICS

By shrinking our world into a village of 100 people, students use proportional thinking to reckon with large numbers and should practice their fluency in going between relative and absolute amounts, in order to maintain the complexity and significance of the problems they are identifying and discussing. Students can use many forms of representing these data and should be encouraged to view data visualizations as another conjugate within mathematical language, interconnected with the symbolic and numeric. Discussions should explore how certain kinds of representations make visible certain features of the given contextual situation, which might lend itself to particular inferences and meaning making. Students can also begin to unpack the significant and nuanced discussion about objectivity and subjectivity as it relates to mathematics and statistics. You should introduce or reintroduce the concepts of counting, sampling, and bias and demonstrate appropriate ways to question data without undermining the complexity of the task. In particular, you should assist students to approach data both without dismissive skepticism akin to these data being made up and without an undue idealism about the origins of these data. Students should discuss complex statistical concepts like precision and accuracy, alongside bias and subjectivity.

Resources and Materials

- Book: *If the World Were a Village: A Book About the World's People* by D. J. Smith (2011)

- Website: 100 People, "A World Portrait" (https://www.100people.org/statistics-100-people/)

  Note: On the 100 People website, "gender" is the first item discussed but it is reported as a binary using sex markers of male and female. This is problematic because it doesn't unpack the diversity of gender (or sex). This is a great opportunity for critical literacy (the same issue happens in most videos in the chart below). Thus, one suggestion is to start with the book and then follow up with the videos listed in Teacher Resource 1.

- Teacher Resource 1: *Additional Video Resources With Characteristics Mentioned in Each*

LESSON 1 FACILITATION

Learn About the World

Launch (20 minutes)

- Introduce the idea of shrinking the world's population down to a village of 100 people that represents it by reading page 7 of *If the World Were a Village* by David J. Smith. Use a globe or map to provide context if necessary. After reading the introduction, have students predict:

  + *How many of these 100 people do you think would live in the United States or North America?*

  + *How many of these 100 people would speak English?*

- You may also engage students in wondering about other characteristics of the world's village. These predictions are meant to serve as prompts for critical literacy and critical quantitative literacy. They cause students to grapple with what they think about the world that might not be reflected in reality or in data. This process is about revealing assumptions that we develop without a clear source other than our personal interactions and relationships and how those assumptions should be challenged in the face of evidence.

- In addition to or instead of reading *If the World Were a Village*, the teacher can also use the video resources listed in Table 1 to examine the world population. Table 1 identifies the social characteristics mentioned in each video weblink.

Table 1. Characteristics Mentioned in Each Video Link

Link	Where from	Languages	Gender	Age	Religion	Shelter	Food	Water	Toilet	Literacy	Education	Income/ Wealth	Electricity	Cell Phone	Internet	Health
If the World Were a Village of 100 People: https://youtu.be/FtYjUv2x65g	X	X														
The 100 People Project: An Introduction: https://www.100people.org/the-100-people-project-an-introduction/	X		X		X			X		X	X		X			
If the World Were a Village of 100 People (2019 Edition): https://youtu.be/aLjlFoOTJlY	X			X	X	X	X	X			X	X	X			
If The World Was 100 People (Jay Shetty): https://youtu.be/LXqOd5noN8g	X	X	X	X	X	X		X		X	X	X	X	X	X	
If The World Were 100 People (GOOD Data): https://youtu.be/QFrqTFRy-LU	X	X	X	X	X	X	X	X		X	X	X		X	X	X
What If Only 100 People Existed on Earth? https://youtu.be/UbffuGZHeRO	X	X	X		X	X		X	X	X	X	X		X	X	X
If the World Was Only 100 People (Knovva): https://youtu.be/A3nIIBT9ACg	X	X	X	X	X	X	X	X	X	X	X			X	X	X

FIELD TESTER REFLECTION

I worked with third graders and I was able to introduce the concept and talk about the world's current population. Using the idea of the Three-Act Tasks, I asked them to make brave low (L), just right (J), and high (H) predictions of how many people are from the U.S. They needed to justify their answers. Here is one of my students' predictions:

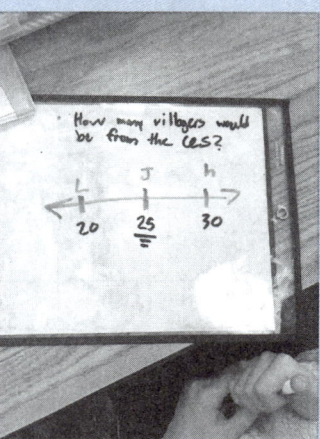

Student A explains estimates https://qrs.ly/svdounf

Student B explains estimates https://qrs.ly/pjdounj

Student C explains estimates https://qrs.ly/e2dounl

I also modified the task slightly, next asking my students: *If each person in the village actually represents 77 million people, how many people are from the U.S? What about the U.S. and Canada?* Here is some student work around these questions.

The United States Only

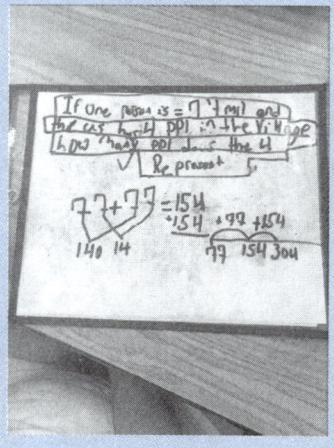

Student D's calculations for the U.S. https://qrs.ly/hgdmz0z

To read a QR code, you must have a smartphone or tablet with a camera. We recommend that you download a QR code reader app that is made specifically for your phone or tablet brand.

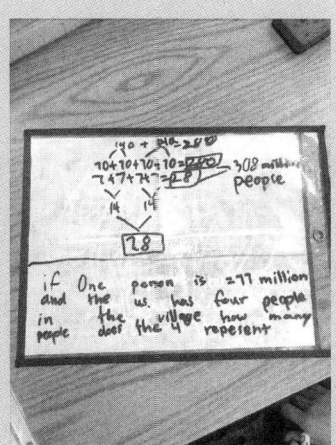

Student F's calculations
for the U.S.

The United States and Canada

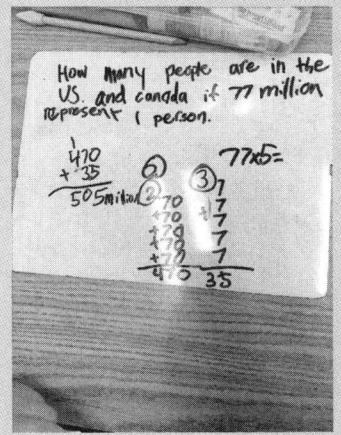

Student G's calculations
for the U.S. and Canada

Student H's calculations
for the U.S. and Canada
https://qrs.ly/zadmz10

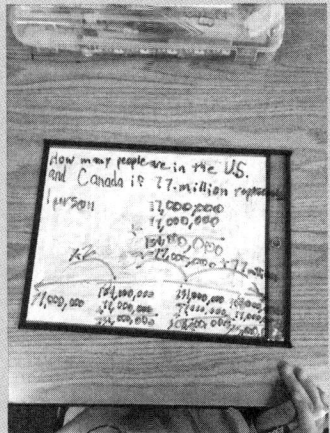

Student I's calculations for
the U.S. and Canada

Source: Ruth Freedman-Finch

- Continue reading the book or viewing a video (see Table 1) to explore the characteristics of the village. (We typically focus on reading the pages about "nationalities," languages, and religions because of the way that the data are presented, but you can also read other pages if children are interested.)

- Explore the languages data with the students. For example, you could ask: *How could we represent this data (e.g., with Unifix cubes)?*

- Also ask students what they notice and wonder about the data. Often, children will make some of the observations noted in Figure 1.

Figure 1. Student Observations About Data

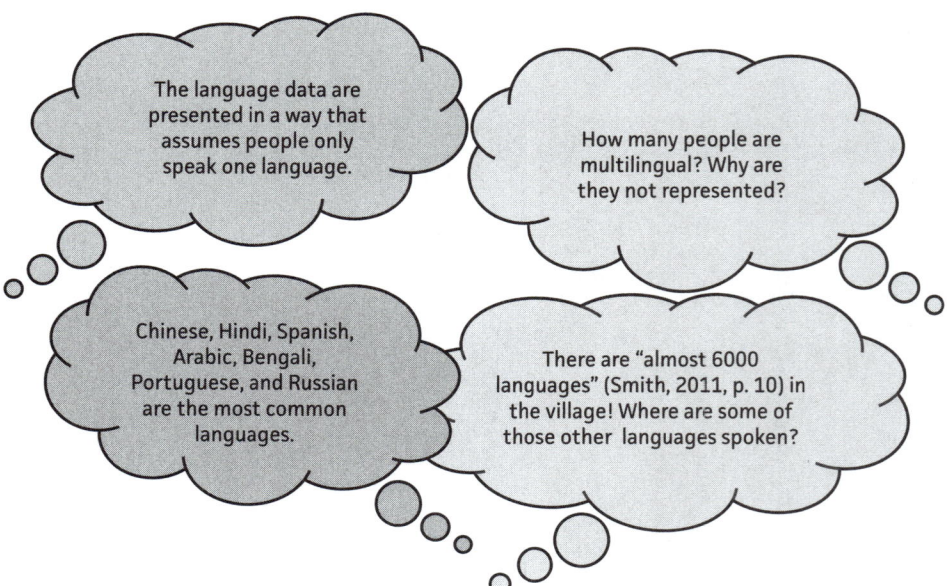

The language data are presented in a way that assumes people only speak one language.

How many people are multilingual? Why are they not represented?

Chinese, Hindi, Spanish, Arabic, Bengali, Portuguese, and Russian are the most common languages.

There are "almost 6000 languages" (Smith, 2011, p. 10) in the village! Where are some of those other languages spoken?

Explore Part 1 (45+ minutes)

TEACHER REFLECTION

I recommend that teachers let the conversations and discussion breathe. Because the point of the lesson is for students to make sense of the complexity of broad topics and related global (or local) issues, it would be counterproductive for the teacher to be the primary driver of the sensemaking. This means that for every group of students, there will likely be different interpretations or different topics that resonate. I spend much of my time listening and taking up students' offered ideas.

—Lynette Guzmán

- Now, students will explore data from other pages in the book and create their own representations of the data on posters to share with their classmates.

- Brainstorm with students multiple ways of representing data including conventional (e.g., conventional ways such as bar charts, pie charts, and pictograms) and unconventional ways such as hundreds charts (10×10 grids).

- Assign groups of four students to explore various topics. (As mentioned earlier, we typically ask students to first explore the religions and the nationalities data because of how the data are presented.)

- Provide independent think time for students to explore the topic and sketch various representations for the topic. Then have small groups share their ideas and decide on several representations to include on a joint poster. On the joint poster, students should include titles, labels, and keys and should try to show the connections between and among the different representations.

TEACHER NOTES

- Prompt students to think about whether another person would understand what they are communicating just by looking at their representations (without the students adding any spoken clarification).

- With respect to learning about objectivity and subjectivity, have students reflect on what aspects of the data are preserved regardless of the representations (e.g., *If you are making a bar chart or pie chart, do the underlying numbers being represented change? What does change?*).

- The fidelity of the data is not subject to personal wishes to communicate a particular conclusion.

- After all posters are created, have students participate in a gallery review and present their findings so everyone can learn about each group's ideas.

- One important aspect of the teaching and learning of statistics and data is the role of representations as communicators. As the gallery review proceeds, have students reflect on what representations communicate the best.

- Ask: *What do different representations communicate? Does the pie chart communicate something different than the bar graph?*

In an effort to use more inclusive language, we use the term *gallery review* for what is commonly referred to as a "gallery walk."

Summarize Part 1 (10 minutes)

- Have students discuss in small groups and then as a whole group:

   + *How do your different representations communicate the data differently?*

LESSON 2 FACILITATION

Learn About Your Local Community Within the World

Launch Lesson 2 (10 minutes)

- Introduce students to data about their country, state, city, town, school district, or school and share with them that they will explore characteristics of this data, similar to how they explored data from *If the World Were a Village*.

- Find data about a context that you and your students are interested in. In our experience, students are very interested in their communities, their cities and towns, and their schools and districts. We are unable to ensure that adequate data exist for localized contexts. In our experience, useful country-level data do exist, and we have found it productive to have students sift through data for their country as a part of the conversation about local contexts. These conversations can launch discussions about generating data at the school or town level.

- Examples of possible existing data to explore include available school district data, language data from individual school sites, state online databases (e.g., California state data from Ed-Data, https://www.ed-data.org/), or country data (e.g., the CIA World Factbook, https://www.cia.gov/the-world-factbook/).

- Have students explore several of the characteristics they explored earlier if the data are available. If not, brainstorm people or organizations they could contact about obtaining the information or discuss whether or not the information is even gathered.

Explore Part 2 (45+ minutes)

- Complete a similar process with these new data as you did for the data from *If the World Were a Village*. Then, use your representations to make comparisons to the world data. Ask students what they notice and wonder about the data, as follows:

  + Ask students to identify the three categories that are the most different between the more local context and the world data (e.g., *Name three ways in which the world and the country are significantly different. How do you know?*)

  + Ask students to identify the three categories that are the most similar between the more local context and the world data (e.g., *Name three ways in which the world and the country are very similar. How do you know?*)

  + Ask students what category they contest the measuring for the most (e.g., *Are there any categories or data that you think are not accurate? How do you know?*)

+ Ask students to assess the validity of data and to become self-aware of the relative strengths and weaknesses of their intuition and expertly generated data (e.g., *How would you rate your level of confidence in each of the categories' data? How do you know if you are correct?*)

Summarize/Act III (20 minutes)

- The following prompts can be used at the end of Lesson 1 or 2, or both:

    + *What do you notice about the representations? What do you wonder?*

    + *What is still confusing?*

    + *What about these data surprised you the most?*

    + *What did not surprise you?*

    + *What questions do you have about these data or the issues related to these data?*

- The two following questions are especially important to get at the critical literacy part of the lesson:

    + *How do your **different** representations communicate the data **differently**?*

    + *What assumptions went into the presentation of the data in the book?*

- In a closing circle reflection, have the students discuss in small groups and then as a whole group:

    + *As we engaged in this lesson learning about the global and local community, what do you appreciate about working with one another here in our classroom community?*

TAKING ACTION

Once students explore the diversity of their local context, they often are curious about resources and supports for different peoples. Teachers can engage students in using knowledge they gain to take action. For example, when students in one second- and third-grade classroom found out that there were a large number of students and families who spoke Spanish and Chinese but their school only had a Spanish Family Liaison working at the school, they wrote letters to the principal to find out why and if there was any way to get support for the Chinese-speaking families in the district.

You may also support students to identify people (e.g., family members, politicians) or organizations (e.g., their own school) that may hold assumptions about who is represented in different spaces and to think about calling in or calling out those assumptions with data.

COMMUNICATING WITH STAKEHOLDERS

You may want to send a letter to parents/caregivers and/or administrators. Here is an example:

We will be reading a children's book and connecting mathematics lessons to the world. The book, If the World Were a Village, *shrinks the world's population down to a village of 100 people, allowing the class to explore how various mathematical representations can be used to describe the world and our local situation within the world.*

ONLINE RESOURCE

 Available for download at **resources.corwin/TMSJ-UpperElementary**

Link	Where from	Languages	Gender	Age	Religion	Shelter	Food	Water	Toilet	Literacy	Education	Income/Wealth	Electricity	Cell Phone	Internet	Health
If the World Were a Village of 100 People: https://youtu.be/FtYjUv2x65g	x	x														
The 100 People Project: An Introduction: https://www.100people.org/the-100-people-project-an-introduction/	x		x		x			x		x	x		x			
If the World Were a Village of 100 People (2019 Edition): https://youtu.be/aLjlFoOTJlY	x			x	x	x	x	x			x	x	x			
If The World Was 100 People (Jay Shetty): https://youtu.be/LXqOd5noN8g	x	x	x		x	x		x		x	x	x	x	x	x	
If The World Were 100 People (GOOD Data): https://youtu.be/QFrqTFRy-LU	x	x	x	x	x	x		x		x	x	x		x	x	x
What If Only 100 People Existed on Earth? https://youtu.be/UbffuGZHeRO	x	x	x	x	x	x	x	x	x	x	x	x		x	x	x
If the World Was Only 100 People (Knovva): https://youtu.be/A3nIlBT9ACg	x	x	x	x	x	x	x	x	x	x	x			x	x	x

Teacher Resource 1: Additional Video Resources With Characteristics Mentioned in Each

ABOUT THE AUTHORS

We are a group of mathematics teacher educators who, in the past, each worked on creating tasks and lessons based on resources such as *If the World Were a Village* (Smith, 2011) or the 100 People World Portrait website (https://www.100people .org/statistics-100-people/) in our respective school settings, from elementary school through higher education. For this book series, we put our heads together to develop a 100 People lesson for the K–2, 3–5, and 6–8 grade bands. While each lesson varies some, our hope is to show how a theme can be followed through in different grade bands.

Lynette Guzmán is a mathematics education scholar who focuses on interrogating limiting discourses about people and their complexity. As a millennial who grew up with the internet, Lynette spends her time thinking about the ways digital platforms lend themselves to content creation, consumption, and remixing to promote particular kinds of discourses.

Jeff Craig is committed to contemplating ethical questions in education. In his teaching, he reconciles ethical questioning against a backdrop of so-called Wicked Problems in education, which prioritizes depth in education as it relates to students as members of communities and societies. Jeff finds this lesson compelling because it engages students as both global and local thinkers who use mathematical and statistical techniques to understand their world and their positions within it.

Eva Thanheiser is a professor of mathematics education in the Mathematics Department at Portland State University. Eva focuses her scholarship on teaching mathematics for social justice, connecting mathematics to social and political issues, and anti-bias mathematics education.

Mary Candace Raygoza is a STEMinist teacher educator at Saint Mary's College of California. Her scholarship explores teaching mathematics for social justice and critical, justice-oriented, anti-racist teacher education. Mary is a former high school mathematics teacher in East Los Angeles.

Courtney Koestler is a proud former public school teacher and currently serves as the Director of the OHIO Center for Equity in Mathematics and Science in the Patton College of Education at Ohio University. Their work centers on critical literacy and critical pedagogies in early childhood and elementary education.

- I feel connected to other people and know how to talk, work, and play with others even when we are different or when we disagree. (Diversity 9)

- I will work with my friends and family to make our school and community fair for everyone, and we will work hard and cooperate in order to achieve our goals. (Action 20)

MATHEMATICS CONCEPTS

- Understand decimal notation for fractions and compare decimal fractions.

- Apply and extend previous understandings of multiplication and division.

- Perform operations with multidigit whole numbers and with decimals to hundredths.

MATHEMATICS PRACTICES

- Make sense of problems and persevere in solving them.

- Reasoning abstractly and quantitatively.

LESSON 7.3 SINGLE-USE PLASTICS

Carolee Hurtado

SOCIAL JUSTICE CONNECTION

This lesson engages students in understanding the magnitude of our plastic usage and its impact on people, animals, oceans, and landfills. Students may not be aware of the Great Pacific Garbage Patch and thus may not realize that trash/plastic often makes it into the ocean. This creates pollution as well as dangers to ocean animals. Students are highlighted as change makers who have been brave to stand up for what they believe in and get the community to collectively work toward a cleaner future. This lesson was created with our local context in mind: Ventura County is adjacent to the Pacific Ocean, and California has instituted plastic bans. This lesson also shares the stories of Milo Cress, Molly Steer, and Shelby O'Neill, who are likely of similar age to your students, and their action to eliminate the use of plastic straws.

TEACHER NOTE

While this lesson was designed to raise awareness of plastics in our community, we should also encourage our students to think about when single-use plastics support and promote accessibility, thus reducing ableism. To learn more about this, please visit the Iowa Law Review article "Water, Water Everywhere, But Not a Straw to Drink: How the Americans with Disabilities Act Serves as a Limitation on Plastic Straw Bans" (https://bit.ly/3oEDGOY).

DEEP AND RICH MATHEMATICS

Students use estimated data on plastic usage, waste, and ingestion to understand large numbers. The mathematical concepts addressed in the lesson include performing whole-number operations, understanding decimal notations, comparing decimals, multiplying whole numbers by a decimal or fraction, and estimating.

Resources and Materials
- Teacher Resource 1: *Single-Use Plastics* (PowerPoint slides)
- Book: *Sip the Straw* by Sam Keck Scott and Woody Heffern (optional)

- Video: "How Much Plastic Is in the Ocean?" from *PBS* (https://to.pbs.org/3DHlRTF)

- Video: "Straw No More," from TEDx (https://bit.ly/3DGCleI)

  + Website: Straw No More (https://www.strawnomore.org)

  + Article: "This 10-year-old girl is ending plastic straw pollution," from 1millionwomen.com, June 6, 2018 (https://bit.ly/3ye3yof)

  + Article: "Shelby O'Neil targets plastic straws in Girl Scout project," by Zoe Papadakis, *Newsmax*, March 19, 2022 (https://bit.ly/36LDebt)

- Website: United Nations Environment Programme, "Beat Plastic Pollution" (https://bit.ly/3oFvq1d)

- Website: National Conference of State Legislatures, "State Plastic Bag Legislation" (https://bit.ly/31KPtTl)

- Website: Footprint Foundation, "Single-Use Plastic Legislation: U.S. Bans at a Glance" (https://bit.ly/3GzETxl)

- Website: Earthday, "Fact Sheet on Plastic Bag Data" (https://bit.ly/3lUbZQr)

- Article: "Reducing plastic as a family is easy: Here's how," by Allyson Shaw, *National Geographic*, June 4, 2018 (https://on.natgeo.com/3dDaGkv)

Optional Resources or Background

- Movie: *A Plastic Ocean* (https://imdb.to/3oFTeSz)

- Website: NRDC, "Single-Use Plastics 101" (https://on.nrdc.org/3dHyCD2)

- Instructional activities were informed by the following articles on microplastics in our immediate environment (what we eat, drink, and breathe).

  + "Most dust motes in our homes are unsafe microplastics, and the Saharan dust cloud may bring more," by Jeff McMahon, *Forbes*, June 28, 2020 (https://bit.ly/3DLtY1r)

  + "Revealed: plastic ingestion by people could be equating to a credit card a week," by WWF-Australia, June 12, 2019 (https://bit.ly/3rR67LD)

  + "How to eat less plastic," by Kevin Loria, *Consumer Reports*, April 30, 2020 (https://bit.ly/3NWhkT8)

+ "The average person eats thousands of plastic particles every year, study finds," by Sarah Gibbens, *National Geographic*, June 5, 2019 (https://on.natgeo.com/3oFVskT)

+ "We're all eating a credit card's worth of PLASTIC each week: Shocking graphics reveal how many millions of sesame-seed size microplastics humans ingest over a month, a year, a decade and a lifetime," by Jonathan Chadwick, *Daily Mail Online*, December 30, 2019 (https://bit.ly/3oDqzh6)

LESSON 1: FACILITATION

Should Single-Use Plastic Items Be Banned in California?

Launch (30 minutes)

- Using the PowerPoint slides for this lesson (Teacher Resource 1), begin by asking students what they know about plastic use, recycling, and plastic bans. Have a brief discussion to learn what students know, as well as what questions they might have.

- Share with students recent information on California's plastic bans (on plastic straws and carryout bags) as shown in the following image (slide 4). (For information on legislation in other states, see websites in the *Resources and Materials*.)

In November 2016, California voters approved Proposition 87, the state-wide Single-Use Carryout Bag Ban. As a result, most grocery stores will no longer be able to provide single-use plastic carryout bags for their customers. Instead, they may provide a reusable or recycled bag for 10 cents.	Passed in November 2018, California Assembly Bill 1884 prohibits full-service restaurants from providing single-use plastic straws unless they are requested.

- Ask: *Do you think California's plastic bans are a good idea?* Have students turn and talk to a partner about their thinking.

- Optional: Read the children's book *Sip the Straw* by Sam Keck Scott and Woody Heffern (slide 5). This book provides a story of a straw's "life" told from the perspective of the straw, starting with the excitement of being used to help a person enjoy their drink to being tossed in the trash, to going out to the Great Pacific Garbage Patch, to washing up onshore. The story introduces students to the enormity of the amount of trash and plastics being dumped into the oceans.

- As a class, watch the PBS video, "How Much Plastic is In the Ocean?" (slide 6) (https://to.pbs.org/3DHlRTF)

- Then ask: *What do you notice? What do you wonder?* (slide 7). Possible student responses might include the following:

  + *Plastic is often designed to be used just once.*

  + *How much is 8,000,000 tons of plastic waste?*

  + *Is there plastic in our food? Do we eat plastic?*

  + *If Americans use 500 million straws per day, what does that say about how many straws one person uses in a day?*

FIELD TESTER NOTE

We spent a lot of time thinking about the question, *If Americans use 500 million straws per day, how many straws per day does the average person use?* This was a question suggested by my students during the notice and wonder, and they were quite motivated to think through the answer. It was a great opportunity to think about large numbers (why 500 million divided by 300 million is the same as 500 divided by 300), prompted them to search for current data on the population themselves, and led to great discussions of the meaning of the number 1.6 straws per day (both in terms of thinking about the quantity 1.6 but also what an average like that tells us about a *typical* day, rather than literally using 1.6 straws at a time).

- The video notes that Americans use 500 million straws per day. This was actually an estimate calculated by a fourth-grader in 2011! (See Chokshi, 2019.) Since then, other research has been conducted to determine Americans' straw use per day, but estimates vary from 170 million to 390 million. You might hold a discussion with students about why the estimates might be so different. For example, *What counts as a straw? Does the straw on a juice box count? Do coffee stirrers count?*

- Pose mathematics questions related to the PBS video estimate of 500 million straws per day (slide 8), or share more recent estimates of straw use (such as 170 million or 390 million; all show a great impact). Ask:

  + *How much is that?*

  + *PBS reports that Americans use 500 million straws per day. Here are other ways to think about this:*

    - *If we line these straws up, how many times will they wrap around the Earth?* (Answer: 4)

    - *How many straws per day does the average person use?* (Answer: 1.6)

Explore (60 minutes)

- Give students individual think and work time, followed by small-group work for the following mathematics tasks (slide 9):

  + *If the average person uses 1.6 straws per day, how many straws would*

    - *The average person use in 1 week? Do you use this many straws in 1 week?*

    - *Your family use in 1 month?*

    - *Our school community use in 1 year?*

- Lead a class discussion on solutions and strategies. Again, give students individual think and work time, followed by small-group work for the following mathematics tasks:

  + *Based on Milo Cress's calculations of 500 million straws per day, the average person uses 38,000+ straws in their lifetime between the ages of 5 and 65.*** (National Park Service, 2021)

  **You could decide here as a class what estimate you would like to use for the number of straws in one's lifetime, and why.

    + *How many straws would you need for your family?*

    + *How much space would you need to store those straws?*

- Lead a class discussion on solutions and strategies.

- Revisit California's Plastic Bans: Proposition 87 and Assembly Bill 1884 (slides 10 and 11). Pose these questions:

+ *Why do you think California is imposing these bans?*

+ *Do you think these bans help our environment? Why or why not?*

- In small groups, and then in a whole-class discussion, have students discuss the following:

  + *What will you think of when you see single-use plastic items such as bags and straws now? What, if anything, do you think California should do to address the abundance of plastic in our landfills and oceans?*

  + *How did using mathematics help you to think about the impact of plastics in our environment?*

  + *Take Action: What are some things you can do to reduce, reuse, recycle, rethink, repair, or refuse?*

Summarize (30 minutes)

- Ask students what they can do to impact their environment. Record student responses.

- Together, watch the video "Straw No More" with Molly Steer (https://bit.ly/3DGCleI) and consider exploring other resources listed in the *Resources and Materials.*

- Ask students to share their reactions to the video and Molly Steer's Straw No More project (slides 12–14).

- Engage the class in a discussion about the power we have as informed citizens (and as young people) to take action against injustices we are passionate about.

- Engage in a think–ink–pair–share instructional activity (slide 15). Have students individually "think" about what we can do to positively impact our environment and then "ink" (write) their ideas. After a couple of minutes, "pair" students to "share" their ideas with one another for a few minutes. The teacher should listen to pair conversations and select some groups to share their ideas.

- Have students write their responses to the following two questions on sticky notes to place on chart paper. Use different colored sticky notes for the two prompts:

  + *What is one thing you learned today?*

  + *What questions or wonderings do you still have?*

- Consider using students' questions and wonderings to design future lessons or investigations (slides 16 and 17).

LESSON 2: FACILITATION

Plastics in Our Environment

Launch (15 minutes)

- Connect this lesson to the previous lesson by asking students: *What impact do you think plastic might have on you personally?*

- Together, look at United Nations Environment Programme Beat Plastic Pollution website (https://bit.ly/3oFvq1d) to establish the fact that we produce 300 million tons of plastic waste each year worldwide, half of which is for single-use items. That's nearly equivalent to the weight of the entire human population. Ask students about this statistic: *What do you notice? What do you wonder?*

- Have students estimate what the world population is based on this statistic.

- Offer more information for students about plastics:

  + Share data on plastic *use* by industrial sector (e.g., the packaging sector produces 146 million tons of plastic per year; the transportation sector produces 27 million tons per year) and plastic *waste* by industrial sector (e.g., the packaging sector generates 141 million tons of plastic waste; the transportation sector generates 17 million tons of plastic waste) (Ritchie & Roser, 2018).

  + Note that Eco (n.d.) estimates that 300,000,000 tons of plastic is produced annually and that 91% of plastics are never recycled.

  + Recall from the PBS video that plastics don't break down; they break into smaller pieces.

- Ask students to discuss: *Where does all of this plastic go?* (slides 20 and 21). Record student responses.

- You might also share additional information that does not arise from students about where scientists are finding plastics:

  + "Scientists find plastic pollution in the rain and in the air we breathe," by Sean Fleming, *World Economic Forum*, July 31, 2020 (https://bit.ly/3IEukuw)

  + "Decomposing plastic revealed as hidden source of greenhouse gas emissions," by Joe McCarthy and Erica Sánchez, *Global Citizen*, August 2, 2018 (https://bit.ly/3oEQV21)

  + Scientists find plastic in waterways, oceans, and rivers; in the food we eat; in the dust we breathe; and in facial scrubs, clothing, and greenhouse gas emissions (slide 22).

- Give students opportunities to share their reactions to and questions about these findings.

- Pose and discuss with students the question: *Are we really eating plastic?* (slide 23).

- Share slide 24 with students and link to this World Wildlife Fund image (https://bit.ly/3rR67LD). Ask: *What do you notice? What do you wonder?*

Explore (40 minutes)

- Share the following with students: *It is estimated that the average person eats 5 grams of plastic each week* (WWF-Australia, 2019) (slide 25).

- Give students individual think and work time, followed by small-group work for the following mathematics tasks:

    + *How much plastic would an average person eat in 1 month?*

    + *In 1 year?*

    + *In 1 lifetime?*

- Lead a whole-class discussion of student strategies for determining the amount of plastic eaten in either 1 year and/or 1 lifetime (slide 26).

- Following this discussion, share visuals for the amounts just calculated.

- Following this discussion, find and share visual representations for the amounts students calculated.

Summarize (5 minutes)

- For the next 3 days, jot down where you see plastic at school, home, and in your community. *What do you notice? What do you wonder?* (slide 29)

LESSON 3 FACILITATION

Plastics in Our Environment: Looking at Our Data

- Lead a whole-class discussion in which students share where they found plastic at school, home, and in their community (slide 31).

- Optional: Have students create visuals representing the data they collected and then engage in a gallery review to look at one another's visuals.

- Chart student responses from their data collection in two categories: (1) plastics we use and/or play with and (2) potentially ingested plastics.

- Have students sit in small groups of three to four. In their groups, have them discuss the following reflection questions (slide 32):

    + *What did you learn about plastic and its impact on people, animals, and the environment? How did math help you think about plastic?*

    + *Plastic is everywhere! It is hard to imagine our world without plastics. What can you do to minimize the impact of plastics in your own body, for other humans and/or animals, in your environment, or in our world?*

- Conclude with a whole-class discussion in which students share their brainstormed ideas about how they would like to act on their learning.

In an effort to use more inclusive language, we use the term *gallery review* for what is commonly referred to as a "gallery walk."

TAKING ACTION

Encourage students to take social action based on what they have learned in this lesson. Here are some ideas to share with students, but have them generate their own ideas and make a plan of action:

- Share what you learned with your families. Perhaps read an article together at home, such as this one from *National Geographic*: "Reducing plastic as a family is easy: here's how" (https://on.natgeo.com/3dDaGkv).

- Start or improve a recycling program in your school.

- Use a refillable water bottle and/or filter drinking water.

- Avoid using plastic utensils.

- Use reusable bags whenever possible.

- Create a toy library so more children can enjoy toys before they get discarded.

- Stop chewing gum. It is made from synthetic rubber (i.e., plastic).

- Speak up! Teach others what you know.

COMMUNICATING WITH STAKEHOLDERS

Teachers may choose to share the learning goals and outcomes with families, colleagues, and/or administrators before, during, or in closing the lesson.

- **Preview:** Students sometimes question why they need to learn mathematics: When or how can they use what they learn in the real world for things they care about? This lesson engages students in using mathematics to consider issues of social justice and environmental justice by looking at how the usage of single-use plastic items impact landfills, oceans, and marine life. Students will have opportunities to create solutions, using mathematics as a means for understanding the significance or magnitude of the issue and developing ways they can make positive impacts in their homes and local and/or national community.

- **During the Lesson:** Teachers can invite various stakeholders such as families and administrators into the classroom to listen to small- and whole-group discussions.

- **Closing the Lesson:** Students can share their learning and solutions with other stakeholders. Teachers can invite families, administrators, other classes, and so on into the learning space for small-group presentations and/or a gallery review.

ONLINE RESOURCE

 Available for download at **resources.corwin/TMSJ-UpperElementary**

Single-Use Plastics

▲
Teacher Resource 1: Single-Use Plastics (PowerPoint slides)

ABOUT THE AUTHOR

Carolee Hurtado is a mathematics educator at California State University Channel Islands who is interested in social justice, critical pedagogy, and centering student voice in K–16 classrooms. One of her favorite places to be is on the classroom rug listening to young mathematicians share their thinking.

NEXT STEPS

CHAPTER 8

ADVICE FROM THE FIELD

We believe that the voices of this book's lesson authors can be meaningful to teachers and those interested in teaching mathematics for social justice (TMSJ). Hearing from other teachers who are dedicated to TMSJ can offer insight and extra enthusiasm for implementing lessons in this book and creating your own social justice mathematics lessons (SJMLs; Chapter 9).

We asked the lesson authors to share some of their experiences, challenges, advice, and efforts to sustain themselves and their communities to persist. Specifically, we asked these questions:

- What has been most valuable about your (and your students') experiences when implementing a TMSJ lesson or lessons?

- What challenges have you been confronted with in the classroom, school, or community when implementing TMSJ lessons, and how did you overcome these challenges? (e.g., facing perceptions of elementary school students being too young or too immature for social justice mathematics)

- What additional bits of advice would you offer anyone implementing this or other TMSJ lessons?

- As a mathematics educator committed to social justice, how do you take care of or support yourself and your community to sustain and persist?

In this chapter, we report their responses, using their words as much as possible. We hope their experiences will provide some inspiration and insight as you set forth to implement the SJMLs in this book or create your own, and enhance your agency while TMSJ.

MOST VALUABLE EXPERIENCES IN TMSJ

Our contributors reported many positive outcomes when implementing TMSJ lessons, including a greater sense of belonging for students, higher levels of engagement, better understanding of their worlds, critical thinking skills, and connecting with others' perspectives. One critical aspect of TMSJ lessons is their potential for making the mathematics classroom an inclusive space for learners. This, combined with the subject matter, can also strengthen student engagement.

I think the most valuable part of the experience for students is the sense of belonging and worth that they experience. In a social justice mathematics classroom, all students are valued, and each one has their own unique voice. It is a safe space where students feel comfortable discussing, sharing, and expanding their mathematical thinking. As a teacher, it is exciting to take a step back and watch the students as they experience these lessons. Often students have an increased ownership of learning shown by high motivation, interest, and work ethic.

— Sarah Ivey

The lessons in this book and those that you create can be a powerful vehicle for helping students make sense of their lives. Including opportunities for students to take action can support them in developing new ways of thinking about mathematics that go beyond typical school justifications to include fighting for a better world.

The most valuable part of my experiences when implementing TMSJ lessons is the engagement from students in seeing that the classroom is a space where they can talk about and try to make sense of their lived experiences. TMSJ reminds students that mathematics is a tool for liberation, not merely a means of getting into college or having a STEM-related career; there are things you can do right now with your mathematics knowledge.

— Evan M. Taylor

In addition to helping students make sense of their own lives, TMSJ lessons can help students develop a deeper appreciation of real-world issues. Strong social justice lessons use mathematics to support investigation of an issue in our world, but they also create a space for meaningful analysis and discussion of that issue. Including space for both helps students understand the significance of real-world topics that they might not have seen before.

I think the most valuable experience from implementing a TMSJ lesson is the conversation that we have afterwards. It is a truly rewarding experience to see students engage in critical, analytical conversations about the world around them and see them become passionate about things that once seemed so mundane, like pay wages. These lessons truly stick with students; I have found the topics circle back over and over again throughout the year.

— Trisha Huynh

FOSTERING A JUSTICE-ORIENTED CLASSROOM

Creating a justice-oriented classroom involves more than the use of one or even several TMSJ lessons. It begins with creating a safe and caring environment with your students in which using mathematics to make sense of the world is the norm.

> *TMSJ is more than integrating a topic about a social injustice into the classroom; it also involves how you establish an environment and interactions with and amongst your students.*
>
> —John W. Staley

> *For me, the purpose of math is not simply applying numbers and formulas to the real world, but on how numbers and equations can help me make sense of the world. Everything we do each day involves math, from figuring out how many clothes can fit in the washing machine to deciding how early to make dinner so that it will be ready on time. Similarly, when approaching broader social issues, math allows us to consider the why and how of the problem and develop a clear strategy for improvement. I encourage you to help your students use this framework for thinking about math, and then completing TMSJ lessons simply makes sense, rather than being an aberration.*
>
> —Rebecca Ellis

Social justice mathematics teachers also work to center their students' lives and concerns in their lessons. This includes following their lead on how to allow lessons to develop. It also involves recognizing that students are already grappling with injustices in their lives and we have a responsibility to understand how they are making sense of those experiences.

> *Let the students' thoughts guide the lessons. There is no need to follow all parts of the lessons. I trust that each teacher knows each of their students.*
>
> —Jennifer Park

> *Our goal, as educators, should be to help each and every student see their mathematical prowess as a tool they can draw on to enact change, both in their personal lives and in the broader community. As adults, we often fail to realize that all students, no matter their age, wrestle with issues of social justice, but sometimes they are not the same issues that we face. Children notice things, want to share their experiences, and feel empowered when we entrust them to explore important issues and propose change. The challenge, for us as educators, is choosing tasks that are meaningful to them.*
>
> —Hyunyi Jung and Megan Wickstrom

PLANNING FOR AND RESPONDING TO CHALLENGES

While TMSJ can be deeply rewarding for teachers and students, it can also raise unique challenges. As we discuss more in Chapter 9, building relationships within and beyond the classroom is a foundational component to TMSJ. These lessons depend on an effective classroom community and on communication with families, especially if there are topics that may be uncomfortable or traumatic for students.

All of these lessons depend on extensive and ongoing relationship building in classroom communities. They cannot be just a lesson that is thrown into the curriculum. Participating in these lessons with children also benefits from communication with families, again ongoing, but also specifically if there are topics that may be sensitive. There need to be ways for children to opt out if they become distressed, and other equally powerful parallel lessons should be available.

—Emma Gargroetzi

A common theme in the advice from our contributors was on the importance of reconceptualizing mathematics teaching and learning. Because of the complexity of the issues involved, TMSJ lessons may not follow a traditional lesson format and instead must make space for student-led discussions. It is also not realistic to expect complete mastery or "perfection" related to complex issues—fostering a safe space that allows for messiness and evolution in thinking is a more appropriate goal.

Any social justice lesson demands time, and might not follow a structured implementation. In my experience, TMSJ should be guided by classroom discussions and should be student-led.

—Debasmita Basu

Learning can be quite a vulnerable experience, especially when navigating topics that have space for worldviews to be challenged. What has been important to me is recognizing that in the learning process, perfection should not be the primary standard; it's more than okay to have messy or incomplete understandings. It's important for me to attend to how students are feeling throughout the process and to promote a safe space where there is no fear of being in progress with your developing understandings.

—Lynette Guzmán

Finally, reconceptualizing "what counts" as an effective mathematics lesson may also require sometimes challenging conversations with colleagues, administrators, and families. Having evidence of the value of TMSJ lessons for students and their engagement, as well as clear connections to mathematics, can help with these conversations.

> *One particular challenge I've had is advocating to include social justice lessons in my school. Colleagues expect me to stick to the curriculum to make sure that the standards are being met. I've been able to confront this by teaching TMSJ lessons anyway and showing my school that they are worth the extra time; these lessons invigorate students because they are related to the real world and they promote a deeper understanding of mathematical concepts.*
>
> —Trisha Huynh

> *Sometimes there is a bit of fugitive practice in teaching these lessons as there is a culture of complicity in school spaces where when you step towards freedom it alarms others, but at the same time it can encourage them to do the same. Many believe some students to be too young to talk about or engage in social justice, but do not carry this same critique when it comes to their lived experiences. I try to call out these inconsistencies, whereas if a child is young enough to experience white supremacy they are also old enough to begin to understand it so that they can disrupt and dismantle white supremacy using mathematics.*
>
> —Evan M. Taylor

THE IMPORTANCE OF COMMUNITY

A theme that comes up repeatedly in research on TMSJ and in the feedback from our contributors is the importance of engaging with supportive communities of social justice–oriented educators and activists. There are many types of communities you may connect with, including the local community and activist groups, colleagues in your building, teachers you meet at conferences, educators you meet online, and supportive friend groups. Some of these groups will be invaluable in helping you identify resources and select lessons that work for and matter to your students.

> *Get to know the community and create bonds with people! Learn the stories of others so that you can find ways to support them!*
>
> —Bethany Chan

I join events organized by social justice communities and keep learning from one another.

—Ho-Chieh Lin

Communication is the key. I believe that TMSJ lessons cannot be implemented in isolation. Listening to other people's experiences and communicating with them can provide educators an insight into different injustices that have plagued our society for ages.

—Debasmita Basu

Community can also help you deepen your own thinking, about social justice topics in our world and about how students will approach them. Opportunities to collaborate with others can help you anticipate students' thinking, better understand diverse perspectives, and refine your lessons.

Find a colleague or a group of colleagues who will do it as well. Talk to each other in advance. What are you anxious about? What are you excited about? How do you plan to modify what is in the book to be directly responsive to your own students? Learn from each other in advance. Run ideas by each other to get some feedback. Then, talk again after the fact. How did it go? What was exciting? What forms of brilliance did this lesson allow you to see in your learners that you don't see in other more traditional lessons? What do you want to do next? Then make goals for your next steps and plan to check in. Encourage each other and hold each other accountable to continue exploring this work.

—Emma Gargroetzi

I stay connected to others who are pushing in on this work. Formal and informal groups provide opportunities to refresh, renew, and recharge. This is needed as the work to Change the Narrative is real.

—John W. Staley

Finally, we encourage you to build a supportive network of friends and colleagues to support you in this challenging, but worthwhile work.

I have good friends who also share my belief that teaching is a work of social justice. Our discussions inspire me to continue my work!

—Trisha Huynh

Community. I surround myself with a community of educators, organizers, professors, and people who love me as I was, as I am, and as I will be. I remember that this work is intergenerational and that I am just doing my part with the time I am allotted to daily try to cause less harm than I did the day before. And naps, I take lots of naps unapologetically.

—Evan M. Taylor

CONCLUSION

While using mathematics to understand and respond to injustice can be challenging, it is also deeply rewarding for teachers and students. We encourage you to start, whether big or small, to find communities that can challenge and support you, and to remember the reasons why we do this work.

The most valuable aspect has been the relevance to students' daily experiences in their community. It is great to be able to give them opportunities to see connections between real-world occurrences and math!

—Joanne Baltazar Vakil

It is a powerful experience to acknowledge your perceptions of the world and see those challenged. When confronted with information that might not seem to fit with your existing worldview, the learning comes into play when you try to make sense of how it can make sense or be better understood. This often provides opportunities for impactful experiences in the classroom.

—Lynette Guzmán

CREATING SOCIAL JUSTICE MATHEMATICS LESSONS FOR YOUR OWN CLASSROOM

CHAPTER
9

We appreciate your interest in teaching mathematics for social justice (TMSJ), and we deeply value and respect the work of the educators that submitted their work for this volume as examples of what it can look like in upper elementary contexts. This chapter is intended to provide some final suggestions for getting started with planning your own social justice mathematics lessons (SJMLs). When teachers begin asking students about what is important to them, there are many relevant, salient social justice issues that can be incorporated into mathematics lessons.

PAUSE AND REFLECT

Before engaging students with SJMLs, it is critical to consider your own social identities and their impact on student learning. How do you identify and position yourself? How do your multiple identities (e.g., race and ethnicity, gender, sexual identity, socioeconomic status, disability status, nationality, languages spoken) connect to your students and relate to the social justice topic under investigation? Often, our own worldviews drive our decisions. Beginning with self-reflection ensures our classroom interactions model the justice work we hope our students and these mathematics lessons will allow us to do as we engage in the world.

Here are a few questions to guide your reflection in your lesson design process:

- What knowledge (aside from the mathematics) and worldview are assumed?

- Whose experiences are reflected or not included?

- How could I ensure the lesson context and my pedagogical decisions reflect a wider number of windows and mirrors for my students to honor and leverage the diversity of social identities in the classroom and school community and also support the social justice and mathematical learning goals in a particular lesson?

- What are ways I can learn more about my students' multidimensional and intersectional identities and draw on student and community strengths, knowledge, and interests?

GETTING STARTED

If you are new to TMSJ, it may seem overwhelming to begin creating SJMLs. Even for more experienced teachers, this kind of teaching is hard work. As discussed in earlier chapters, there can be many tensions in developing these kinds of lessons. However, there are many positive, affirming possibilities as well, and we believe that elementary students deserve time and space to explore issues relevant and important to them and to use mathematics as a tool to understand and create change. The following list represents eights steps that we recommend you consider in developing SJMLs, followed by elaboration of each step:

1. Learn About Your Students and Their World

2. Learn About the Relevant Social Injustices

3. Identify the Mathematics

4. Establish Your Goals

5. Determine How You Will Assess Your Goals

6. Create a Social Justice Question for the Lesson

7. Design the Student Resources for the Investigation

8. Plan for Action and Reflection

Step 1: Learn About Your Students and Their World

How can I listen to the other, how can I hold a dialogue, if I can only listen to myself, if I can only see myself, if nothing or no one other than myself can touch me or move me? If while humble, one does undermine oneself or accepts humiliation, one is always ready to teach and to learn. Humility helps me avoid being entrenched in the circuit of my own truth.

—Freire (2005, p. 72)

Freire charges educators to develop the quality of humility in order to truly listen and dialogue with the students, families, and communities in which we serve. Humility is central to developing SJMLs. Earlier in this book, we discussed that effective TMSJ in upper elementary settings builds from the interests, passions, questions, and concerns students have about their school, community, nation, or world—what Freire (1970/2000) refers to as generative themes. The students' questions or concerns create the opportunity for problem-posing pedagogy, an instructional process that engages students in noticing and wondering about and taking action in their world. In an ideal setting, these generative themes would drive curriculum, and teachers could match their mathematical goals to those themes, rooted in concerns about social injustice.

Begin by listening and learning more about your students and the local community. Engage in asset mapping of the community to identify the strengths and needs of their physical and social environment. Get to know the community; shop at the local markets and vendors; and learn about the local community organizations, community history, and local governance to help you in identifying potential partners and resources to support your lesson implementation. Listen to students and engage in collective investigation of the local community as a part of instruction. Investigations with our students, their parents/caregivers, and local community members will inform the ideas, hopes, doubts, fears, and questions they have about their family, friends, school, community, nation, or world, serving as generative themes for your lessons. This may be students' first time engaging in asset mapping of their own community. The following prompts can be helpful to get individual students, families, or the whole class to identify strengths and resources as well as interests and concerns:

- What activities do you and your family do at home (e.g., celebrations, gardening, cooking)?

- What are the places that are important to you and your family in the neighborhood?

- Where do you go in your neighborhood? Why?

- What places do you avoid? Why do you avoid them?

- Imagine someone was new to the community/area. What kinds of things would you tell them they could do or go see?

- What is happening in our town/city/community/area/school that you're proud of? (Choose one for a prompt.)

- What is happening in our town/city/community/area/school that you would like to change? (Choose one for a prompt.)

- What are some things going on in the country or world that you're excited about?

- What are some things going on in the country or world that you wish you could change?

Step 2: Learn About the Relevant Social Injustices

Elementary teachers have always been leaders in PreK–12 education when learning about contexts that were engaging and relevant to students and how to incorporate these in tasks, activities, and lessons. However, when TMSJ it is also important to learn all you can about the issue of injustice that you and/or your students want to explore.

Many issues will have local or community meaning or impact. Your students, their parents/caregivers, and their family members will be able to identify social issues and potential injustices right in your area that might directly involve them, their families, and/or community members. Tap into the expertise of families and consider reaching out to community members and organizations that focus on responding to the topic or issue. Attend a meeting and scour their website. Find a member of the organization to speak with. Mathematics often is or can be used to understand or respond to injustice (for example, in understanding and collecting data, analyzing statistics, or creating a timeline of events). The better you understand local issues and connect to local organizations, the easier it will be to identify mathematical connections.

You can also review local news and media outlets to learn what news items students may be discussing in school, with friends, or in their out-of-school lives. Whether local, national, or even international news sites, online media are often good sources to expose students to broader issues at the local, state, national, and international levels; as such, these resources can also be used as part of your lesson. Many news outlets post stories to their internet sites, which allows you to share articles and videos with your students. However, it's also important to note that all sources of information come with bias, whether it be news or other forms of media or information – even community groups. (For more on developing critical literacy and examining messages and meaning with young children, see the work of Vivian Vasquez, 2016.)

We've encouraged you to collaborate with school colleagues to develop allies in your work. This is an excellent opportunity to involve the school librarian or other teachers (1) to learn more about bias in information sources and (2) to learn how to counteract and balance the information you receive as well as present to students.

Our first step still asks you to learn about the questions and concerns of the students and the community with whom you work. However, rather than immediately pursuing these questions in conjunction with your students, it is oftentimes important to learn more about the topic on your own first, reflect on your own understanding and perspectives around the topic, and learn from others in the community and from additional sources to develop broader perspectives. Our intent is that you've identified a few potential social injustices from which to build a SJML aligned with the mathematics content you must teach.

Step 3: Identify the Mathematics

When TMSJ, the task, activity, or lesson you design must allow you to connect with mathematics that is important for students to use and learn in meaningful ways. But what mathematics is involved in the potential issues you've learned about from your students in Steps 1 and 2?

We have a few strategies that have worked for us. One strategy is to begin with the social injustice context, starting with an artifact for students to analyze. For example, in Lesson 5.5 (*Exploring Maskmatics! Sociocultural and Environmental Concerns With Disposable Masks During COVID-19*), the teacher shares an image with students and asks *What do you notice? What do you wonder?* Students will naturally connect to their own and community experiences and mathematics can be used to deepen the discussion through prompts such as these: *How many disposable masks are thrown away each second? Minute? Hour? Day?* These questions allow students to develop a sense of scale of millions, starting with smaller quantities and building from a familiar context.

Another is to build from children's literature. Many children's books are rich with mathematical investigations. For example, in Lesson 5.1 (*Families Matter*), the teacher reads aloud *My Friends and Me* to draw attention to different family structures available, prompting students to examine the representation of families and family structures in their set of school books and to create data displays to explain their results.

Another strategy is to begin with the content. However, instead of focusing on a single standard, which can sometimes be too narrow, we recommend focusing on the cluster name, which groups together several related standards. These are used in many state standards as well as the Common Core State Standards; for example, the third-grade standards on Operations and Algebraic Thinking include "Represent and solve problems involving multiplication and division" (National Governors Association for Best Practices, Council of Chief State School Officers, 2010, p. 23).

A fourth approach that has worked well for us is to look at a few of our favorite well-designed tasks or task sequences, such as those in this volume and in this series. Often, teachers are gifted at looking at examples and using the context of the other tasks to provide insight into how they could be modified or replaced with the context of a different generative theme. Thus, much of the rich mathematical development is already in place, and we are modifying the context to match the questions or concerns students have about a social issue or injustice.

Step 4: Establish Your Goals

When developing goals for class explorations, set both mathematics and social justice goals. The mathematics goals should focus on both mathematics content standards as well as the mathematical practices and processes in which students will engage. Here you should draw upon your local expectations and resources, such as district or state mathematics learning objectives or standards. You can plan your social justice goals with regard to identity, diversity, justice, or action using the Learning for Justice (2016) Social Justice Standards for guidance (see Appendix D). Goals related to social justice might also be related to developing, supporting, and challenging students' understanding about an issue. Finally, you might also consider how this is situated in your classroom related to other content areas and/or routines already established.

We encourage you to write only a small number of learning outcomes for each lesson. While it can be easy to see many connections mathematically and to the Social Justice Standards, when you articulate too many goals it can be difficult to maintain focus on teaching or learning any of them. Consider setting only one mathematical and one social justice learning outcome depending on the size and scope of the task, activity, lesson, or unit. This will allow you to maintain focus on what students should be learning.

Step 5: Determine How You Will Assess Your Goals

To help maintain your focus on student learning outcomes, you will need to identify how students' understanding of and development in both the mathematics and social issues will be assessed in your TMSJ task, activity, lesson, or unit and how these relate to your goals. Formative assessment will be central in ascertaining what students already know about the social issue at the start of a lesson, and it will support you in connecting on and building on these understandings throughout the exploration. Include brief check-ins throughout to allow you and the students to assess one or more of your learning goals. Plan to pose social justice and mathematics questions to assess and advance students' thinking.

Reflection activities at the end of the experiences promote student self-reflection on their learning and allow you to gauge whether you are reaching goals for the whole class as well as for individual students. Later (in Step 8), we discuss having a plan for action and reflection, which is another way you can gauge the extent to which students achieve both the mathematical and social justice learning goals in their final product.

Step 6: Create a Social Justice Question for the Lesson

Develop a guiding question for your lesson based on the social justice issue at hand. This can come from the students themselves or it can be one you have crafted based on what you know about student questions, concerns, and the topic itself. Choosing a question should be both interesting and powerful as a guide for students to explore, understand, and respond to the issue.

Remember, one aim for your SJML is for your students to see how mathematics can be used to help them understand the issue and consider action. The question(s) used to drive the lesson might make that explicit or be grounded in a mathematical approach to better understand the social injustice. Sometimes it may seem like the mathematics is in the foreground or background as well as integrated with other content areas, depending on the lesson.

Step 7: Design the Student Resources for the Investigation

Developing the resources to support student learning of the mathematical and social justice learning outcomes—through exploration driven by the social justice question—forms the heart of the investigation.

Identify the resources students might draw upon to learn more about the social issue—likely a few of the same you used to better understand the issue. Materials with the following qualities are most effective. We suggest that the resources should

- Be written specifically for an upper elementary student audience;

- Include various kinds of media (print resources, including text and photos, short clips of video), and you might even include guest speakers that could join the class in person or via video conference;

- Center voices that represent those most impacted (and therefore most knowledgeable) by the social issue; and

- Identify common assumptions and misunderstandings about the context, and be prepared to offer opposing perspectives on a controversial issue.

You will also need to structure the mathematical investigation of the lesson. In our work, we have accomplished this in one of two ways. Sometimes, we rely on our experience with many different curricula and similar mathematical tasks so that we can identify a series of prompts or tasks that allow students to better understand the content through deepening their understanding of mathematics. Structure the lesson to allow students to explore, reason, create, justify, generalize, and/or apply, rather than recall, compute, or calculate before interpreting the meaning in the context of the lesson. The latter approach not only creates more shallow mathematical understanding, but it also can cause students to lose ownership of the investigation process.

The most success we have had designing the investigation portion of the SJML is to gain inspiration from or modify strong lessons we've used in the past. Using the structure and sequence of questions from an established, quality lesson is often the most reliable way to ensure a lesson that promotes reasoning, problem solving, and quality engagement of your students.

Ensure that when the tasks are implemented, there is plenty of opportunity for student-to-student discourse. Often, it is most effective to write this into the student resources, or at least into the teacher notes. Opportunities for discourse can be accomplished by assigning pairs or small groups to debrief a resource, collaborate on problem solving, plan for a presentation and discussion of ideas with classmates, and more. Revisit the discourse section of Chapter 3 (especially Figures 3.7 and 3.8) to help you plan these opportunities into your investigation.

As a final bit of advice: If you haven't already been developing the lessons with others, consider asking a colleague and/or community leaders with expertise in this area for feedback on the first draft of the student resources. You might even try the activity with a small group of students and solicit their feedback.

Step 8: Plan for Action and Reflection

The final element of an effective SJML is to ensure opportunities for reflection and action. Consider the following questions: What individual, small-group, and whole-class opportunities will you provide for students to reflect on the lesson and consider possible actions they can take to address an injustice? How will students share what they have learned about the social injustice? How might they use the mathematical knowledge and skills developed in this lesson?

The lessons in this book provide different examples of how teachers engaged and supported students in reflection and action. The Action domain of the Learning for Justice Social Justice Standards also provides insight into different ways we might think of action in upper elementary classrooms. For example, students might commit to stand up to the exclusion, prejudice, or bias surrounding the lesson topic. You might support them in deciding what an appropriate and effective response might be.

Better yet, ask students how they want to take action.

IN SUMMARY

If you have read to this point in the book, you have already begun your journey to integrate social injustice into your classroom, and we encourage you to use and share this book with others as you engage in all phases of preparing to teach a SJML. Our hope is that this book will inspire you and others to build on students' agency as learners and doers of mathematics who can use mathematics as a tool to explore, understand, and respond to issues of social injustice.

Each of us has a role to play in shaping the future of the mathematics education community, so connect with others who embrace TMSJ, who are using SJMLs in their classrooms, and who are continuing to learn about the topics discussed in this book, who are taking action to make a difference, and who are members of a TMSJ community willing to share and learn together. We invite you to connect with us through Facebook at Math Lessons to Explore Social Injustice and Twitter at @UESJMath.

APPENDIX A: ADDITIONAL RESOURCES

Additional Examples of Social Justice Mathematics Lessons

Better World Ed. *I Am Taniya: Creating Confidence.* https://betterworlded.org/stories/i-am-taniya-creating-confidence

Esmonde, I., & Caswell, B. (2010). Teaching mathematics for social justice in multicultural, multilingual elementary classrooms. *Canadian Journal of Science, Mathematics and Technology Education, 10*(3), 244–254. https://doi.org/10.1080/14926156.2010.504485

Felton-Koestler, M. D., Simic-Muller, K., & Menéndez, J. M. (2017). *Reflecting the world: A guide to incorporating equity in mathematics teacher education.* Information Age.

Gutstein, E., & Peterson, B. (Eds.). (2013). *Rethinking mathematics: Teaching social justice by the numbers* (2nd ed.). Rethinking Schools.

Guzmán, L. D., & Craig, J. (2019). The world in your pocket: Digital media as invitations for transdisciplinary inquiry in mathematics classrooms. *Occasional Paper Series, 2019*(41), 6. https://educate.bankstreet.edu/cgi/viewcontent.cgi?article=1274&context=occasional-paper-series

Kinser, K. STEM by the Numbers. *Learning for Justice.* https://www.learningforjustice.org/classroom-resources/lessons/stem-by-the-number

Learning for Justice. *Calculating the Poverty Line.* https://www.learningforjustice.org/classroom-resources/lessons/calculating-the-poverty-line

Learning for Justice. *Cultural Relevancy in the Cafeteria.* https://www.learningforjustice.org/classroom-resources/lessons/cultural-relevancy-in-the-cafeteria

Learning for Justice. *What Do Halloween Costumes Say?* https://www.learningforjustice.org/classroom-resources/lessons/what-do-halloween-costumes-say

Ontario Ministry of Education. (n.d.). *Teaching mathematics through a social justice lens.* https://thelearningexchange.ca/projects/teaching-mathematicsthrough-a-social-justice-lens/

Osler, J. (2007). Math and social justice. *RadicalMath.* http://www.radicalmath.org/math-social-justice

Raygoza, M. C. (2016). Striving toward transformational resistance: Youth participatory action research in the mathematics classroom. *Journal of Urban Mathematics Education, 9*(2), 122–152.

Robertson Program. (2020, January). *Five ways to explore social justice through mathematics.* https://wordpress.oise.utoronto.ca/roberton/2020/01/13/five-ways-to-explore-social-justice-through-mathematics

Stocker, D. (2006). *Maththatmatters: A teacher resource linking math and social justice.* CCPA Education Project Canadian Centre for Policy Alternatives.

Stocker, D. (2017). *Maththatmatters 2: A teacher resource linking math and social justice.* CCPA Education Project. Canadian Centre for Policy Alternatives.

Thanheiser, E., & Koestler, C. (2021). If the world were a village: Learning mathematics while learning about the world. *Mathematics Teacher Educator, 9*(3), 202–228. https://doi.org/10.5951/MTE.2020.0021

Turner, B. (2019, January). *Open secrets in first-grade math: Teaching about white supremacy on American currency.* https://www.learningforjustice.org/maganize/open-secrets-in-firstgrade-math-teaching-about-white-supremacy-on-american-currency

Turner, E. E., & Font Strawhun, B. T. (2007). Posing problems that matter: Investigating school overcrowding. *Teaching Children Mathematics, 13*(9), 457–463. https://doi.org/10.5951/TCM.13.9.0457

Learn More About Elements of TMSJ

A Pathway to Equitable Math Instruction. (2020). *Dismantling racism in mathematics instruction: Exercises for educators to reflect on their own biases to transform instructional practice.* https://equitablemath.org/wp-content/uploads/sites/2/2020/11/1_STRIDE1.pdf

Aguirre, J., Mayfield-Ingram, K., & Martin, D. B. (2013). *The impact of identity in K–8 mathematics: Rethinking equity-based practices.* NCTM.

Benjamin Banneker Association. (2017). *Implementing a social justice curriculum: Practices to support the participation and success of African-American students in mathematics.* http://bbamath.org/wp-content/uploads/2017/11/BBA-Social-Justice-Position-Paper_Final.pdf

Chval, K., Chavez, O., Pomerenke, S., & Reams, K. (2009). *Mathematics for every student: Responding to diversity, Grades PreK–5.* NCTM.

Drake, C., Aguirre, J. M., Bartell, T. G., Foote, M. Q., Roth McDuffie, A., & Turner, E. E. (2015). *TeachMath learning modules for K–8 mathematics methods courses—Teachers Empowered to Advance Change in Mathematics Project.* http://www.teachmath.info

Felton, M. D. (2010). Is math politically neutral? *Teaching Children Mathematics, 17*(2), 60–63.

Gewertz, C. (2020). Teaching math through a social justice lens. *EducationWeek.* https://www.edweek.org/teaching-learning/teaching-math-through-a-social-justice-lens/2020/12

Gutstein, E. (2006). *Reading and writing the world with mathematics*. Routledge

Huinker, D., & Bill, V. (2017). *Taking action: Implementing effective mathematics teaching practices in K–Grade 5*. NCTM.

Murrey, D. (2008). Making numbers count. Retrieved from https://www.learning-forjustice.org/magazine/spring-2008/making-numbers-count

NCSM & TODOS: Mathematics for ALL. (2016). *Mathematics education through the lens of social justice: Acknowledgment, actions, and accountability*. https://www.todos-math.org/socialjustice

Stinson, D. W., & Wager, A. A. (2012). Teaching mathematics for social justice: Conversations with educators. NCTM.

Su, F., & Jackson, C. (2020). *Mathematics for human flourishing*. Yale University Press.

Swoveland, M. (2013). On 'that's so gay' and learning math. *Learning for Justice*. https://www.learningforjustice.org/magazine/on-thats-so-gay-and-learning-math

Wiki for ideas for Social Justice in Math. Retrieved from https://sites.google.com/site/mathandsocialjustice/curriculum-resources/wiki-for-ideas-for-social-justice-in-math

Yeh, C. (2021). Mathematics in context: The pedagogy of liberation. *Learning for Justice*. https://www.learningforjustice.org/magazine/spring-2021/mathematics-in-context-the-pedagogy-of-liberation

Yeh, C., Ellis, M. W., & Hurtado, C. K. (2017). *Reimagining the mathematics classroom*. NCTM.

Teaching for Social Justice

Abolitionist Teaching Network: https://abolitionistteachingnetwork.org/

Blake, C. (n.d.). Teaching social justice in theory and practice. *Resilient Educator*. https://resilienteducator.com/classroom-resources/teaching-social-justice

Christensen, L., Hansen, M., Peterson, B., Barbian, E., & Watson, D. (2012). *Rethinking elementary education*. Rethinking Schools.

Facing History and Ourselves: https://www.facinghistory.org

Gonzalez, J. (2016, February 14). A collection of resources for teaching social justice. https://www.cultofpedagogy.com/social-justice-resources

Great Lakes Equity Center: https://greatlakesequity.org

Kohl, H. (n.d.). Teaching for Social Justice 15.2. *Rethinking Schools*. https://rethinkingschools.org/articles/teaching-for-social-justice

Learning for Justice: https://www.learningforjustice.org

New York Collective of Radical Educators. *Revolutionizing the classroom: Transforming mainstream curriculum into social justice teaching.* http://www.nycore.org/newsite/wp-content/uploads/revolutionizingtheclassroom.pdf

Peterson, B. (n.d.). Teaching for Social Justice 8.3. *Rethinking Schools.* https://rethinkingschools.org/articles/teaching-for-social-justice-8–3

Picower, B. Six elements of social justice pedagogy. *Using Their Words.* http://www.usingtheirwords.org/6elements/

Rethinking Schools: https://www.rethinkingschools.org

Southern Illinois University, Edwardsville. (n.d.). *Resource list for including social justice issues in curricula.* https://www.siue.edu/education/about/diversity/resources.shtml

UCLA Ethnic Studies Teacher Resources: http://www.teachethnicstudies.org/

Communities Interested in Teaching Mathematics for Social Justice

Abolitionist Teaching Network: https://abolitionistteachingnetwork.org/

Black Teacher Project: https://www.blackteacherproject.org/

Education for Liberation Network: https://www.edliberation.org/

Educators for Social Justice & Inclusive Teaching Practices—Facebook: https://www.facebook.com/groups/1136557443074208

Equity and Social Justice in Mathematics Education Facebook: https://www.facebook.com/groups/178344199241717

Free Minds Free People: https://fmfp.org/

Graphs in the World

   Instagram: @graphsintheworld

   Facebook: https://www.facebook.com/graphsintheworld

National Network of Teacher Activist Groups: https://teacheractivists.org/

Nepantla Teachers Community: https://nepantlateachers.wixsite.com/website

New York Collective of Radical Educators (NYCoRE)—Facebook: https://www.facebook.com/NYCoRE

People's Education Movement Bay Area

   Online: https://peoplesedbayarea.wordpress.com/

   Instagram: @peoplesbayarea

People's Education Movement LA

Online: https://peoplesed.weebly.com/

Instagram: @peoples_ed

Social Justice Math—Twitter: @socjusticemath

Teachers 4 Social Justice Bay Area: https://t4sj.org/

Teachers for Social Justice Chicago: http://www.teachersforjustice.org/

Teaching for Change

Facebook: https://www.facebook.com/TeachingforChange

Online: https://www.teachingforchange.org

Teaching on Days After: Dialogue & Resources for Educating Toward Justice Facebook: https://www.facebook.com/groups/teachingondaysafter

Teaching Social Justice Resource Exchange—Facebook: https://www.facebook.com/groups/teachaboutjustice

Washington State Ethnic Studies Now: https://waethnicstudies.com/

Witness for Peace solidarity collective: https://www.solidaritycollective.org

Resources for Building Your Own Social Justice Mathematics Lessons

A Little Stats: https://alittlestats.blogspot.com/p/data-sources.html

Bigelow, B. (n.d.) Videos with a global conscience. *Rethinking Schools*. https://rethinkingschools.org/books/rethinking-globalization/videos-with-a-global-conscience

Borderlinks: https://www.borderlinks.org/

Data for Black Lives: http://d4bl.org/

DATAJUSTICE project: https://datajusticeproject.net

EdGap: http://edgap.org

Gallup: https://www.gallup.com/

Gapminder: https://www.gapminder.org

GLSEN: https://www.glsen.org

Mathematical Modeling with Cultural and Community Contexts: https://sites.google.com/a/uw.edu/dr-julia-aguirre/research/mathematical-modeling-with-cultural-and-community-contexts-m2c3

Smith, D. J. (2009). *If America were a village: A book about the people of the United States.* Kids Can Press.

Smith, D. J. (2011). *If the world were a village: A book about the world's people* (2nd ed.). Kids Can Press

Social Justice Books: https://socialjusticebooks.org/store/

U.S. Census Bureau: https://www.census.gov/en.html

What's Going on in This Graph, *New York Times*: https://www.nytimes.com/column/whats-going-on-in-this-graph

 Available for download at
resources.corwin/TMSJ-UpperElementary

APPENDIX B: LESSON RESOURCES

Lesson 5.1: *Families Matter*

- Book: *My Friends and Me* by Stephanie Stansbie
- Video: Ms. Katie reading, *My Friends and Me* (https://bit.ly/3y4o6zd) or an alternative video Ali Ayars reading this book (https://bit.ly/3Dv3Wj7)
- Worksheet 1: *Family Type Tally Sheet*
- Teacher Resource 1: *Sample Letter*

Lesson 5.2: *Playground Prejudice*

- Video: "Get Along Monsters: Don't Call Me Names" (https://bit.ly/3IsOWpG)
- Teacher Resource 1: *Survey Tool*
- Worksheet 1: *Task 1—Notice and Wonder Data* from the Gay, Lesbian and Straight Education Network (GLSEN) survey, "Playgrounds and Prejudice: Elementary School Climate in the United States" (https://bit.ly/3ItWMPF)
- Worksheet 2: *Task 2—Task Cards*
- Worksheet 3: *Task 3*

Lesson 5.3: *Who Appears in Billboards?*

- Worksheet 1: *Making Sense of Your Research*

Lesson 5.4: *Family Story Problems*

- Book: *Children's Mathematics: Cognitively Guided Instruction* by T. Carpenter and colleagues (2015)
- Worksheet 1: *Sketch and Solve Template*

Lesson 5.5: *Exploring Maskmatics! Sociocultural and Environmental Concerns With Disposable Masks During COVID-19*

- Article and Graph: "More Americans say they are regularly wearing masks in stores and other businesses," by Stefanie Kramer, *Pew Center Research*, August 27, 2020 (https://pewrsr.ch/31ua2U3)
- Article and Graph: "Wear a mask? Yes, always wear a mask," *Institute for Health Metrics and Evaluation*, June 18, 2020 (https://bit.ly/339K8Ft)
- Video: "Pandemic Pollution: Disposable Masks, Gloves Are Saving Lives But Ruining the Environment," by Stephanie Sy and Lorna Baldwin, *PBS News*, April 22, 2021 (https://to.pbs.org/3IwbPs7)
- Article: "Where did 5,500 tonnes of plastic waste from masks end up?" by Jenny Yeh, *Greenpeace*, August 14, 2020 (https://bit.ly/3rJLelx)
- Article: "Three million masks every minute: how Covid-19 is choking the planet," *The Straits Times*, January 9, 2021 (https://bit.ly/3ow2pF2)
- Website: Centers for Disease Control and Prevention (CDC) stance on face coverings (https://bit.ly/339xshV)
- Worksheet 1: *Mask Disposal Data Collection Sheet*

(Continued)

(Continued)

Optional

- Article: "Living with facemasks: How to stow them, reuse disposables and more," by Emily Chung, *CBC News*, August 6, 2020 (https://bit.ly/3rIJzg0)
- Article: "All about the Coronavirus," *Junior Scholastic*, September 18, 2020 (https://bit.ly/3IvytB6)
- Article: "Face masks: What the data say," by Lynne Peeples, *Nature*, October 6, 2020 (https://go.nature.com/3dtq2b7)
- Article: "Face masks: New solutions to reduce their negative impact on the environment," by Abigail Saltmarsh, *Medical Expo*, January 6, 2021 (https://bit.ly/3EunCVw)
- Article: "How to stop discarded face masks from polluting the planet," by Laura Parker, *National Geographic*, April 14, 2021 (https://on.natgeo.com/3dAhGyh)
- Family Resource: "Coronavirus (COVID-19): Kids and Masks," Nemours Children's Health (https://bit.ly/31Ewf1p)
- Poster: "All About Masks," *Sesame Street* (https://bit.ly/3y6XwWb)
- Poster: "Be a Mask Hero," *Junior Scholastic* (https://bit.ly/3GsVa7f)
- Tutorial: "How to Make a Facemask," *CBC Kids News* (https://bit.ly/3drK85J)
- Video: "Covid-19: Does Your Kid Really Need a Mask?" from *NOVA* (https://to.pbs.org/3y7ywOO)
- Website: Mayo Clinic Network, "Benefits of Kids Wearing Masks *in* School." (https://mayocl.in/3rLrxKg)

Lesson 5.6: *Challenging Ableist Assumptions in Mathematics Problems*

Picture Books

- *The Bug Girl* by Sophia Spencer
- *Emmanuel's Dream: The True Story of Emmanuel Ofosu Yeboah* by Laurie Ann Thompson and Sean Qualls
- *Hello Goodbye Dog* by Maria Gianferrari
- *I Am Not a Label* by Cerrie Burnell
- *A Kids Book About Disabilities* by Kristine Napper
- *Mama Zooms* by Jane Cowen-Fletcher
- *Rescue and Jessica: A Life-Changing Friendship* by Jessica Kensky and Patrick Downes
- *Terry Fox and Me* by Mary Beth Leatherdale
- *What Happened to You?* by James Catchpole and Karen George

Chapter Books

These may be more appropriate for older grades.

- *Braced* by Alyson Gerber
- *Roll With It* by Jamie Sumner
- *Intersectional Allies: We Make Room for All* by Chelsea Johnson, LaToya Council, and Carolyn Choi

Reference Books for You

- *Critical Literacy Across the K–6 Curriculum* by Vivian Maria Vasquez

Additional Resources

- Article: "How to talk to your kid about disabilities," by Caroline Bologna, *Huffington Post*, March 1, 2021 (https://bit.ly/32Svi68)
- Lesson: Learning for Justice, "Picturing Accessibility: Art, Activism and Physical Disabilities" (https://bit.ly/3oeLVkC)
- Lesson: Learning for Justice, "What Is Ableism?" (https://bit.ly/3oe0kh9)
- Lesson: Learning for Justice, "What Is a Disability?" (https://bit.ly/3lrFFUG)

Lesson 6.1: *"Tu lucha es mi lucha": Mathematics for Movement Building*

- Book: *Journey for Justice: The Life of Larry Itliong* by Dawn B. Mabalon and Gayle Romasanta
- Video: "AAPI Civil Rights Heroes—Asian Americans Advancing Justice," from the Zinn Education Project (https://bit.ly/32F9r1N) (Scroll down the page to find the video.)
- Article: "Farmworker Wages in California," by Philip Martin and Daniel Costa, *Economic Policy Institute Working Economics Blog*, March 21, 2017 (https://bit.ly/3rMdS5E)
- Article: "Mapping UFW strikes, boycotts, and farm worker actions 1965-1975," by Katie Anastas, Civil Rights and Labor History Consortium (https://bit.ly/3EBYHiO)
- PBS Documentary: *The Farm Worker Movement* (https://bit.ly/2ZIjzWq)
- Worksheet 1: *Place Value Chart* (Lesson 2 for decimal understanding and representation)
- Worksheet 2: *Making Sense of the Wages* (Lesson 2 for understanding what workers earned)
- Teacher Resource 1: *Resources for the Gallery Review* (Lesson 3)

Lesson 6.2: *Exploring Equitable Pay for Work*

- Worksheet 1: *Data Suggesting Wage Inequalities*
- Article: "Gender pay gap in U.S. held steady in 2020," by Amanda Barroso and Anna Brown, *Pew Research Center* (https://pewrsr.ch/3y3K6dJ)
- Article: "Quick facts about the gender wage gap," *Center for American Progress* (https://ampr.gs/3Ev42lN)
- Article: Survey data from "The state of the gender pay gap in 2021," *Payscale* (https://bit.ly/3lMLMn0)

Websites

- President John F. Kennedy signing the Equal Pay Act (https://bit.ly/3GjMIa0)
- Encountering Racism with Courage and Protest: Protest for Equal Pay (https://bit.ly/3y5tbaC)
- Equal Pay for Native Women (go to http://www.equalpaytoday.org and find the link for Native Women)
- Latina Equal Pay Day (https://www.latinaequalpay.org/about)

Articles

- "The fight for equal pay . . . 40 years on," by Jo Revill, *The Guardian*, May 31, 2008 (https://bit.ly/3oxupYY)
- "1970 Women's Strike for Equality" by Linda Napikoski, *Thought.Co*, February 24, 2019 (https://bit.ly/3rJCQT9)
- "Equal Pay Day: Women rally against wage gap, workplace discrimination," *NBC News*, April 4, 2017 (https://nbcnews.to/3EPHnqZ)
- "Equal Pay Day for Asian American and Pacific Islander Women," from the Feminist Majority Foundation (https://bit.ly/3lsvA3T)
- "Equal Pay for Black Women," from National Women's Law Center (https://bit.ly/3ozVevE)

(Continued)

(*Continued*)

- "Those with disabilities earn 37% less on average; Gap even wider in some states," American Institutes for Research (https://bit.ly/3DuPuaF)
- "The gay and transgender pay gap," by Center for American Progress (https://ampr.gs/3EEVOOk)
- "Before Team USA Women's Hockey won Olympic gold, they won equality off the ice," by Alix Langone, *Money*, February 18, 2018 (https://bit.ly/33bSIJl)
- "The root of WNBA players' argument about gender wage gap," by Adam Grosbard, *The Atlanta Journal Constitution*, June 28, 2018 (https://bit.ly/3IrB3YK)

Optional

- Book: *Joelito's Big Decision/La Fan Decision do Joelito* by Ann Berlak
- Article: "U.S. women's soccer won 4 World Cups. Now can they score equal pay?" by Gretchen Frazee and Yasmeen Alamiri, *PBS News Hour*, July 8, 2019 (https://to.pbs.org/3Ivy1m7)
- Article: "How to bridge that stubborn pay gap," by Claire Cain Miller, *The New York Times*, January 15, 2016 (https://nyti.ms/3Ivvx7u)
- Article: "Actions and Words; They Bring Change" public learning plan, from Learning for Justice (https://bit.ly/3IvcoCF)

Lesson 6.3: *Modeling Library Funding*

- Video: "How America's Public Schools Keep Kids in Poverty," from Kandice Sumner, TED (https://bit.ly/3IMxNh4)
- Teacher Resource 1: *Three Images of Library Books*
- Teacher Resource 2: *Data Tables*

Lesson 6.4: *The Value of a School Lunch*

- Worksheet 1: *Wants vs. Needs*
- Worksheet 2: *What Fraction of Your Recommended Daily Value Is in Our School Lunch?*
- Teacher Resource 1: *Incorporating Decimals via Money Calculations* (Optional Extension Activity)
- Student Resource 1: *What Is the Cost of School Lunch?* (Extension Worksheet)
- Video: "The Cost of Food Insecurity in Schools," from the Health Forward Foundation (https://bit.ly/3rQcw9U)

Lesson 6.5: *More Than an Athlete*

- Worksheet 1: *Roster Selection and Justification Sheet*

Websites

- NBA All-Star Roster (https://www.nba.com/2021-all-star-roster)
- 2021 NBA All-Star Game Guide for background information about the All-Star Game player selection (https://www.nbcsports.com/philadelphia/sixers/nba-all-star-2021-complete-guide-rosters-updated-format-and-more)
- Jr. NBA resource to support understanding of NBA statistics (https://jr.nba.com/how-to-read-a-box-score/)

Videos

- James Harden (https://youtu.be/YeeDMFF9WUk)
- Giannis Antetokounmpo (https://youtu.be/qdgc5yitWHg)

Lesson 6.6: *Your Action Saves Lives: COVID-19 and Systems Thinking*

- Article: "Talking to Children About COVID-19," from *Sesame Street* (https://bit.ly/3DxiwXr)
- Website: The COVID Tracking Project racial demographics for your state (https://bit.ly/3Iu1WLP)
- Student Resource 1: *Decimals Chart*
- Article: "Why COVID-19 Is a Social Justice Issue," by Jai Phillips, *LATogether*, May 15, 2020 (https://bit.ly/3pF20zz)
- Activity: Talking About COVID-19 With Children (https://sesamestreetincommunities.org/activities/talking-about-covid-19-with-children/)

Lesson 7.1: *Water Is Our Right, Water Is Our Responsibility*

- Book: *We Are Water Protectors* by Carole Lindstrom
- Video: "Why Care About Water?" from *National Geographic* (https://bit.ly/3ya2OjG)
- Image from article: "Judge orders Dakota Access Pipeline spill response plan, with tribe's input," by Phil McKenna, *Inside Climate News*, December 5, 2017 (https://bit.ly/3oEjXyN)
- Video: "Dakota Access Pipeline Explained: What Is It, Why Are People Protesting It, and What Happens Next?" by Mythili Sampathkumar, *Independent*, January 24, 2017 (https://bit.ly/3ybZdBH)
- Maps from article: "These maps help fill the gaps on the Dakota Access Pipeline," by Lyndsey Gilpin, *High Country News*, November 5, 2016 (https://bit.ly/3IBPwkY)
- Article: "Oil, water, and steel: The Dakota Access Pipeline," *Earth Justice* (https://bit.ly/31MX1Vv)
- Article: "Keystone Pipeline History," Mark Hefflinger, *Bold Nebraska*, November 7, 2019 (https://bit.ly/3dF4isO)
- Worksheet 1: *Say, Mean, Matter Graphic Organizer*
- Teacher Resource 1: *How Much Water Do We Use?* (PowerPoint slides)

Lesson 7.2: *Upper Elementary Mathematics to Explore People Represented in Our World and Community*

- Book: *If the World Were a Village: A Book About the World's People* by D. J. Smith (2011)
- Website: 100 People, "A World Portrait" (https://www.100people.org/statistics-100-people/)
- Teacher Resource 1: *Additional Video Resources With Characteristics Mentioned in Each*

Lesson 7.3: *Single-Use Plastics*

- Teacher Resource 1: *Single-Use Plastics* (PowerPoint slides)
- Book: *Sip the Straw* by Sam Keck Scott and Woody Heffern (optional)
- Video: "How Much Plastic Is in the Ocean?" from *PBS* (https://to.pbs.org/3DHlRTF)
- Video: "Straw No More," from TEDx (https://bit.ly/3DGClel)
- Website: Straw No More (https://www.strawnomore.org)
- Article: "This 10-year-old girl is ending plastic straw pollution," from 1millionwomen.com, June 6, 2018 (https://bit.ly/3ye3yof)
- Article: "Shelby O'Neil targets plastic straws in Girl Scout project," by Zoe Papadakis, *Newsmax*, March 19, 2022 (https://www.newsmax.com/thewire/shelby-oneil-plastic-straws-girl-scout/2018/06/25/id/868144/?msclkid=8abd1f39a7b711ec9352c28f0680ee12)
- Website: United Nations Environment Programme, "Beat Plastic Pollution" (https://bit.ly/3oFvq1d)
- Website: National Conference of State Legislatures, "State Plastic Bag Legislation" (https://bit.ly/31KPtTl)
- Website: Footprint Foundation, "Single-Use Plastic Legislation: U.S. Bans at a Glance" (https://bit.ly/3GzETxl)
- Website: Earthday, "Fact Sheet on Plastic Bag Data" (https://bit.ly/3IUbZQr)
- Article: "Reducing plastic as a family is easy: Here's how," by Allyson Shaw, *National Geographic*, June 4, 2018 (https://on.natgeo.com/3dDaGkv)

(Continued)

(Continued)

Optional Resources or Background

- Movie: *A Plastic Ocean* (https://imdb.to/3oFTeSz)
- Website: NRDC, "Single-Use Plastics 101" (https://on.nrdc.org/3dHyCD2)
- Instructional activities were informed by the following articles on microplastics in our immediate environment (what we eat, drink, and breathe).
    + "Most dust motes in our homes are unsafe microplastics, and the Saharan dust cloud may bring more," by Jeff McMahon, *Forbes*, June 28, 2020 (https://bit.ly/3DLtY1r)
    + "Revealed: plastic ingestion by people could be equating to a credit card a week," by WWF-Australia, June 12, 2019 (https://bit.ly/3rR67LD)
    + "How to eat less plastic," by Kevin Loria, *Consumer Reports*, April 30, 2020 (https://www.consumerreports.org/health-wellness/how-to-eat-less-plastic-microplastics-in-food-water-a8899165110/)
    + "The average person eats thousands of plastic particles every year, study finds," by Sarah Gibbens, *National Geographic*, June 5, 2019 (https://on.natgeo.com/3oFVskT)
- "We're all eating a credit card's worth of PLASTIC each week: Shocking graphics reveal how many millions of sesame-seed size microplastics humans ingest over a month, a year, a decade and a lifetime," by Jonathan Chadwick, *Daily Mail Online*, December 30, 2019 (https://bit.ly/3oDqzh6)

 Available for download at **resources.corwin/TMSJ-UpperElementary**

APPENDIX C: CATALYZING CHANGE: FIVE MATHEMATICAL CONTENT DOMAINS IN GRADES 3-5

Content Domain	Children Need More Opportunities to . . .
Whole-Number Concepts and Operations	• Develop flexibility in reasoning with number and operation relationships. • Use subitizing activities across the grades to develop quantitative relationships. • Learn basic number combinations through sensemaking, not memorization. • Transition successfully and intentionally from additive to multiplicative thinking.
Fraction Concepts and Operations	• Use unit fractions as the building blocks for developing fraction knowledge. • See fractions as numbers whose magnitude can be represented on a number line. • Use real-world contexts for understanding fraction operations conceptually.
Early Algebraic Concepts and Reasoning	• Develop meaning for the equals sign as stating two expressions have the *same value*. • Discuss observations and intuitions about the properties and behaviors of operations. • Experience algebraic thinking across the mathematics curriculum.
Data Concepts and Statistical Thinking	• Use data to describe the variability of phenomena in their world. • Create data displays to organize, analyze, and communicate information. • Use data distributions to answer questions and pose further questions.
Geometry and Measurement Concepts and Spatial Reasoning	• Develop spatial reasoning as an essential core of mathematical development. • Co-construct meaning of attributes in two- and three-dimensional geometric shapes. • Discuss, understand, and quantify measurable attributes of shapes.

Source: Huinker, D. (2020). *Catalyzing change in early childhood and elementary mathematics: initiating critical conversations*. NCTM.

APPENDIX D: SOCIAL JUSTICE STANDARDS AND TOPICS

	Social Justice Standards (Learning for Justice, 2016)	
	Identity	
1	Students will develop positive social identities based on their membership in multiple groups in society.	
2	Students will develop language and historical and cultural knowledge that affirm and accurately describe their membership in multiple identity group.	
3	Students will recognize that people's multiple identities interact and create unique and complex individuals.	
4	Students will express pride, confidence, and healthy self-esteem without denying the value and dignity of other people.	
5	Students will recognize traits of the dominant culture, their home culture, and other cultures and understand how they negotiate their own identity in multiple spaces.	
	Diversity	
6	Students will express comfort with people who are both similar to and different from them and engage respectfully with all people.	
7	Students will develop language and knowledge to accurately and respectfully describe how people (including themselves) are both similar to and different from each other and others in their identity groups.	
8	Students will respectfully express curiosity about the history and lived experiences of others and will exchange ideas and beliefs in an open-minded way.	
9	Students will respond to diversity by building empathy, respect, understanding, and connection.	
10	Students will examine diversity in social, cultural, political, and historical contexts rather than in ways that are superficial or oversimplified.	
	Justice	
11	Students will recognize stereotypes and relate to people as individuals rather than representatives of groups.	
12	Students will recognize unfairness on the individual level (e.g., biased speech) and injustice at the institutional or systemic level (e.g., discrimination).	
13	Students will analyze the harmful impact of bias and injustice on the world, historically and today.	
14	Students will recognize that power and privilege influence relationships on interpersonal, intergroup, and institutional levels and consider how they have been affected by those dynamics.	

Social Justice Standards (Learning for Justice, 2016)	
15	Students will identify figures, groups, events, and a variety of strategies and philosophies relevant to the history of social justice around the world.
Action	
16	Students will express empathy when people are excluded or mistreated because of their identities and concern when they themselves experience bias.
17	Students will recognize their own responsibility to stand up to exclusion, prejudice, and injustice.
18	Students will speak up with courage and respect when they or someone else has been hurt or wronged by bias.
19	Students will make principled decisions about when and how to take a stand against bias and injustice in their everyday lives and will do so despite negative peer or group pressure.
20	Students will plan and carry out collective action against bias and injustice in the world and will evaluate what strategies are most effective.

Source: Reprinted with permission of Learning for Justice, a project of the Southern Poverty Law Center. https://www.learningforjustice.org/.

Selected Social Justice Topics, Chapters 5–7
Cultural and Global Diversity
Disability/Ableism
Education Justice
Environmental Justice
Food Insecurity
Health Care Justice
LGBTQIA+ Rights

APPENDIX E: LESSONS BY CATALYZING CHANGE MATHEMATICAL CONTENT DOMAINS, SOCIAL JUSTICE OUTCOMES, AND SOCIAL JUSTICE TOPICS

Lessons by Mathematical Content Domain

Mathematical Content Domain	Lesson Title
Whole Numbers and Operations	5.6: *Challenging Ableist Assumptions in Mathematics Problems*
	6.5: *More Than an Athlete*
	7.1: *Water Is Our Right, Water Is Our Responsibility*
	7.2: *Upper Elementary Mathematics to Explore People Represented in Our World and Community*
	7.3: *Single-Use Plastics*
Fraction Concepts and Operations	5.1: *Families Matter*
	5.4: *Family Story Problems*
	6.1: *"Tu lucha es mi lucha": Mathematics for Movement Building*
	6.2: *Exploring Equitable Pay for Work*
	6.3: *Modeling Library Funding*
	6.4: *The Value of a School Lunch*
	6.5: *More Than an Athlete*
	6.6: *Your Action Saves Lives: COVID-19 and Systems Thinking*
	7.3: *Single-Use Plastics*
Data Concepts and Statistical Thinking	5.2: *Playground Prejudice*
	5.3: *Who Appears in Billboards?*
	5.5: *Exploring Maskmatics! Sociocultural and Environmental Concerns With Disposable Masks During COVID-19*
	6.2: *Exploring Equitable Pay for Work*
	6.3: *Modeling Library Funding*
	6.5: *More Than an Athlete*
	7.2: *Upper Elementary Mathematics to Explore People Represented in Our World and Community*

Lessons by Social Justice Outcome, Grades 3–5

Social Justice Outcome, Grades 3–5 (Learning for Justice, 2016)	Lesson Title
Identity	
1 Students will develop positive social identities based on their membership in multiple groups in society.	5.3: *Who Appears in Billboards?*
2 Students will develop language and historical and cultural knowledge that affirm and accurately describe their membership in multiple identity group.	5.4: *Family Story Problems*
5 Students will recognize traits of the dominant culture, their home culture, and other cultures and understand how they negotiate their own identity in multiple spaces.	5.1: *Families Matter* 5.4: *Family Story Problems* 5.5: *Exploring Maskmatics! Sociocultural and Environmental Concerns With Disposable Masks During COVID-19*
Diversity	
6 Students will express comfort with people who are both similar to and different from them and engage respectfully with all people.	5.1: *Families Matter* 5.3: *Who Appears in Billboards?* 5.6: *Challenging Ableist Assumptions in Mathematics Problems*
7 Students will develop language and knowledge to accurately and respectfully describe how people (including themselves) are both similar to and different from each other and others in their identity groups.	5.6: *Challenging Ableist Assumptions in Mathematics Problems* 7.2: *Upper Elementary Mathematics to Explore People Represented in Our World and Community*
8 Students will respectfully express curiosity about the history and lived experiences of others and will exchange ideas and beliefs in an open-minded way	5.4: *Family Story Problems* 6.6: *Your Action Saves Lives: COVID-19 and Systems Thinking* 7.1: *Water Is Our Right, Water Is Our Responsibility* 7.2: *Upper Elementary Mathematics to Explore People Represented in Our World and Community*
9 Students will respond to diversity by building empathy, respect, understanding, and connection.	5.5: *Exploring Maskmatics! Sociocultural and Environmental Concerns With Disposable Masks During COVID-19* 5.6: *Challenging Ableist Assumptions in Mathematics Problems* 7.3: *Single-Use Plastics*
10 Students will examine diversity in social, cultural, political, and historical contexts rather than in ways that are superficial or oversimplified.	6.1: *"Tu lucha es mi lucha": Mathematics for Movement Building* 6.2: *Exploring Equitable Pay for Work*

(Continued)

(Continued)

Social Justice Outcome, Grades 3–5 (Learning for Justice, 2016)	Lesson Title
Justice	
11 Students will recognize stereotypes and relate to people as individuals rather than representatives of groups.	5.4: *Family Story Problems*
12 Students will recognize unfairness on the individual level (e.g., biased speech) and injustice at the institutional or systemic level (e.g., discrimination).	5.1: *Families Matter* 5.3: *Who Appears in Billboards?* 6.1: *"Tu lucha es mi lucha": Mathematics for Movement Building* 6.2: *Exploring Equitable Pay for Work* 6.3: *Modeling Library Funding*
13 Students will analyze the harmful impact of bias and injustice on the world, historically and today	5.2: *Playground Prejudice* 6.2: *Exploring Equitable Pay for Work*
14 Students will recognize that power and privilege influence relationships on interpersonal, intergroup, and institutional levels and consider how they have been affected by those dynamics.	5.5: *Exploring Maskmatics! Sociocultural and Environmental Concerns With Disposable Masks During COVID-19* 6.3: *Modeling Library Funding* 6.4: *The Value of a School Lunch* 6.6: *Your Action Saves Lives: COVID-19 and Systems Thinking* 7.2: *Upper Elementary Mathematics to Explore People Represented in Our World and Community*
15 Students will identify figures, groups, events, and a variety of strategies and philosophies relevant to the history of social justice around the world.	6.1: *"Tu lucha es mi lucha": Mathematics for Movement Building* 7.1: *Water Is Our Right, Water Is Our Responsibility*
Action	
16 Students will express empathy when people are excluded or mistreated because of their identities and concern when they themselves experience bias.	5.2: *Playground Prejudice*
17 Students will recognize their own responsibility to stand up to exclusion, prejudice, and injustice	5.2: *Playground Prejudice* 5.3: *Who Appears in Billboards?* 6.5: *More Than an Athlete* 7.1: *Water Is Our Right, Water Is Our Responsibility*
18 Students will speak up with courage and respect when they or someone else has been hurt or wronged by bias	6.5: *More Than an Athlete*

Social Justice Outcome, Grades 3–5 (Learning for Justice, 2016)	Lesson Title
19 Students will make principled decisions about when and how to take a stand against bias and injustice in their everyday lives and will do so despite negative peer or group pressure.	6.1: *"Tu lucha es mi lucha": Mathematics for Movement Building* 6.2: *Exploring Equitable Pay for Work* 6.5: *More Than an Athlete*
20 Students will plan and carry out collective action against bias and injustice in the world and will evaluate what strategies are most effective.	5.1: *Families Matter* 5.5: *Exploring Maskmatics! Sociocultural and Environmental Concerns With Disposable Masks During COVID-19* 6.3: *Modeling Library Funding* 6.4: *The Value of a School Lunch* 6.5: *More Than an Athlete* 6.6: *Your Action Saves Lives: COVID-19 and Systems Thinking* 7.3: *Single-Use Plastics*

Lessons by Social Justice Topic

Social Justice Topic	Lesson Title
Bullying and Bias	5.2: *Playground Prejudice*
Cultural, Racial, and Global Diversity	5.3: *Who Appears in Billboards?* 5.4: *Family Story Problems* 7.2: *Upper Elementary Mathematics to Explore People Represented in Our World and Community*
Disability/Ableism	5.6: *Challenging Ableist Assumptions in Mathematics Problems*
Economic Inequality	6.2: *Exploring Equitable Pay for Work*
Education Justice	6.3: *Modeling Library Funding*
Environmental Justice	5.5: *Exploring Maskmatics! Sociocultural and Environmental Concerns With Disposable Masks During COVID-19* 7.1: *Water Is Our Right, Water Is Our Responsibility* 7.3: *Single-Use Plastics*
Food Insecurity	6.4: *The Value of a School Lunch*
Health Care Justice	6.6: *Your Action Saves Lives: COVID-19 and Systems Thinking*
LGBTQIA+ Rights	5.1: *Families Matter*
Race and Ethnicity	5.3: *Who Appears in Billboards?* 6.5: *More Than an Athlete*
Rights and Activism	6.1: *"Tu lucha es mi lucha": Mathematics for Movement Building* 6.5: *More Than an Athlete*

 Available for download at **resources.corwin/TMSJ-UpperElementary**

APPENDIX F: SOCIAL JUSTICE MATHEMATICS LESSON PLANNER

PART I

CONTEXT	
Purpose	
Audience	
Allies	
Timing	

CONTENT	
Mathematics Goal(s)	**Mathematical Content Domain**
Social Justice Topic and Brief Description	
Social Justice Outcome	

WHO	
Resources for Your Learning	Classroom Practices and Norms to Establish Social and Emotional Support
How Your Lesson Supports Students in Recognizing Injustice at Both Individual and Institutional or Systemic Levels	

WHEN
Possible Interdisciplinary Connections

HOW
Instructional Model (e.g., Classroom Routine, Mathematics Task, Three-Act Task)

PART II

Launch/Engagement	
What will the teacher do?	What will students do?

Exploration/Investigation	
What will the teacher do?	What will students do?

Summarize/Discuss	
What will the teacher do?	What will students do?

Taking Action	

Stakeholder Communication	
How is the teacher communicating lesson goals?	How are students communicating their learning?

REFERENCES

Agarwal-Rangnath, R., Yeh, C., & Hsieh, B. (2022, forthcoming). We need to see each other as human: Ethnic studies as a framework for humanizing K–12 education. In T. K. Chapman and N. Hobbel (Eds.), *Social justice pedagogy across the curriculum: The practice of freedom*. Routledge.

Aguirre, J., Benedicto, R., & Brown, K. (2018, April). *TODOS: Social justice mathematics* [Pre-conference workshop]. National Council of Teachers of Mathematics 2018 Annual Meeting & Exposition, , Washington, DC.

Aguirre, J., Mayfield-Ingram, K., & Martin, D. B. (2013). *The impact of identity in K–8 mathematics: Rethinking equity-based practices*. NCTM.

Anti-Defamation League. (2021). *What is anti-bias education?* https://www.adl.org/education/resources/glossary-terms/what-is-anti-bias-education

Artiles, A. J., Klinger, J. K., & Tate, W. F. (2006). Representation of minority students in special education: Complicating traditional explanations. *Educational Researcher, 35*(6), 3–5.

Banks, J. A. (2004). Multicultural education: Historical development, dimensions, and practice. In J. A. Banks & C. A. M. Banks (Eds.), *Handbook of research on multicultural education* (pp. 3–29). Jossey-Bass.

Barajas-Lopez, F., & Larnell, G. V. (2019). Unpacking the links between equitable teaching practices and standards for mathematical practice: Equity for whom and under what conditions? *Journal for Research in Mathematics Education, 50*(4), 349–361. https://doi.org/10.5951/jresematheduc.50.4.0349

Barshay, J. (2018, May 7). 20 judgments a teacher makes in 1 minute and 28 seconds: A researcher says 'micro moments' in the classroom reveal implicit biases, subtle racism and sexism. *The Hechinger Report*. https://hechingerreport.org/20-judgments-a-teacher-makes-in-1-minute-and-28-seconds/

Bartell, T. B. (2013). Learning to teach mathematics for social justice: Negotiating social justice and mathematical goals. *Journal for Research in Mathematics Education, 44*(1), 129–163. https://doi.org/10.5951/jresematheduc.44.1.0129

Bartell, T. B., Wager, A., Edwards, A., Battery, D., Foote, M., & Spencer, J. (2017). Toward a framework for research linking equitable teaching with the standards for mathematical practice. *Journal for Research in Mathematics Education, 48*(1), 7–21. https://doi.org/10.5951/jresematheduc.48.1.0007

Benjamin Banneker Association. (2017). *Implementing a social justice curriculum: Practices to support the participation and success of African-American students in mathematics*. http://bbamath.org/wp-content/uploads/2017/11/BBA-Social-Justice-Position-Paper_Final.pdf

Berry, R. Q., III, Conway IV, B. M., Lawler, B. R., & Staley, J. W. (2020). *High school mathematics lessons to explore, understand, and respond to social injustice*. Corwin.

Bishop, R. S. (1990, Summer). "Mirrors, windows, and sliding glass doors." Originally appeared in *Perspectives: Choosing and Using Books for the Classroom, 6*(3). Accessed November 2014. https://scenicregional.org/wp-content/uploads/2017/08/Mirrors-Windows-and-Sliding-Glass-Doors.pdf

Boaler, J. (2015). *Mathematical mindsets. Unleashing students' potential through creative math. Inspiring messages and innovative teaching.* Wiley.

California Legislative Information. (2000). *California Labor Code Section 510.* https://leginfo.legislature.ca.gov/faces/codes_displaySection.xhtml?lawCode=LAB&sectionNum=510

Carpenter, T. P., Fennema, E., Franke, M. L., Levi, L. & Empson, S.B. (2015). *Children's mathematics: Cognitively guided instruction* (2nd ed.). Heinemann.

CAST. (2018). *Universal Design for Learning Guidelines version 2.2.* https://udlguidelines.cast.org

Center for American Progress. (2020, March 24). *Quick facts about the gender wage gap.* https://www.americanprogress.org/article/quick-facts-gender-wage-gap/

Chokshi, N. (2019, July 19). How a 9-year-old boy shaped a debate on straws. *New York Times.* https://nyti.ms/3oDMW66

Curriculum Corner. (n.d.). *Decimal place value chart.* https://www.thecurriculumcorner.com/thecurriculumcorner456/wp-content/pdf/math/decimals/decimal%20place%20value%20chart.pdf

Cvencek, D., Meltzoff, A. N., & Greenwald, A. G. (2011). Math-gender stereotypes in elementary school children. *Child Development, 82*(3), 766–779. https://doi.org/10.1111/j.1467–8624.2010.01529.x

Delgado, R. (1990). When a story is just a story: Does voice really matter? *Virginia Law Review, 76,* 95–111. https://doi.org/10.2307/1073104

Derman-Sparks, L., Edwards, J. O., & Goins, C. (2020). *Anti-bias education for young children and ourselves* (2nd ed.). National Association for the Education of Young Children.

Deutsch, A. (1944). The first US census of the insane (1840) and its use as pro-slavery propaganda. *Bulletin of the History of Medicine, 15*(5), 469–482. https://www.jstor.org/stable/44446305

Dillard, C. (2019, Fall). Black minds matter: Interrupting school practices that disregard the mental health of Black youth. *Learning for Justice, 63.* https://www.learningforjustice.org/magazine/fall-2019/black-minds-matter

Dingle, M., & Yeh, C. (2021). Mathematics in context: The pedagogy of liberation. *Learning for Justice.* https://www.learningforjustice.org/magazine/spring-2021/mathematics-in-context-the-pedagogy-of-liberation

Dunn, A. H. (2021). *Teaching on days after: Educating for equity in the wake of injustice.* Teachers College Press.

Earth Justice. (n.d.). *Oil, water, and steel: The Dakota Access Pipeline.* https://bit.ly/31MX1Vv

Eco. (n.d.). *The house made of plastic.* https://bit.ly/3lQLOKx

Featherstone, H., Crespo, S., Jilk, L. M., Oslund, J. A., Parks, A. N., & Wood, M. B. (2011). *Smarter together! Collaboration and equity in the elementary math classroom.* NCTM.

Flores, A. (2007). Examining disparities in mathematics education: Achievement gap or opportunity gap? *High School Journal, 91*(1), 29–42. http://www.jstor.org/stable/40367921

Freire, P. (2000). *Pedagogy of the oppressed* (M.B. Ramos, Trans.). Continuum. (Original work published 1970)

Freire, P. (2005). *Teachers as cultural workers: Letters to those who dare teach* (Expanded edition). Routledge.

Galindo, E., & Lee, J. (2018). *Rigor, relevance, and relationships: Making mathematics come alive with project-based learning.* NCTM.

Goodman, M. E. (1952). *Race awareness in young children.* Collier Books.

González, N., Andrade, R., Civil, M., & Moll, L. (2001). Bridging funds of distributed knowledge: Creating zones of practices in mathematics. *Journal of Education for Students Placed at Risk*, 6(1–2), 115–132. https://doi.org/10.1207/S15327671ESPR0601-2_7

González, N., Moll, L. & Amanti, C. (Eds.). (2005). *Funds of knowledge: Theorizing practices in households, communities, and classrooms.* Erlbaum Associates.

Gutstein, E. (2006). *Reading and writing the world with mathematics: Toward a pedagogy for social justice.* Routledge.

Howell, C. (2020, November). To sustain the tough conversations, active listening must be the norm. *Learning for Justice.* https://www.learningforjustice.org/magazine/to-sustain-the-tough-conversations-active-listening-must-be-the-norm

Huinker, D. (2020). *Catalyzing change in early childhood and elementary mathematics.* NCTM.

Jansen, A., Cooper, B., Vascellero, S., & Wandless, P. (2017). Rough-draft talk in mathematics classrooms. *Mathematics Teaching in the Middle School*, 22(5), 304–307. https://doi.org/10.5951/mathteacmiddscho.22.5.0304

Jones, S. P. (2020, Spring). Ending curriculum violence. *Learning for Justice.* https://www.learningforjustice.org/magazine/spring-2020/ending-curriculum-violence

Jost, J. T., & Kay, A. C. (2010). Social justice: History, theory, and research. In S. T. Fiske, D. T. Gilbert, & G. Lindzey (Eds.), *Handbook of social psychology* (pp. 1122–1165). John Wiley & Sons. https://doi.org/10.1002/9780470561119.socpsy002030

Ladson-Billings, G. (2009). *The dreamkeepers: Successful teachers of African American children.* Jossey-Bass. (Original work published 1994)

Ladson-Billings, G. (2019, April). *Are we still solving for X? The pedagogical practices limiting student success in mathematics.* Opening keynote presentation at the National Council of Teachers of Mathematics 2019 Annual Meeting & Exposition, San Diego, CA.

Learning for Justice. (2016). *Social justice standards: The teaching tolerance anti-bias framework.* https://www.tolerance.org/magazine/publications/social-justice-standards

Learning for Justice. (2017). *Civil discourse in the classroom.* https://www.learningforjustice.org/sites/default/files/2017–10/Civil-Discourse-v2-CoverRedesign-Oct2017.pdf

Learning for Justice. (2018). *The teaching tolerance social justice standards: A professional development facilitator guide.* https://www.learningforjustice.org/sites/default/files/2018–11/TT-Social-Justice-Standards-Facilitator-Guide-WEB_0.pdf

Learning for Justice. (2021). *Toolkit for "mathematics in context": The pedagogy of liberation.* Retrieved from: https://www.learningforjustice.org/magazine/spring-2021/toolkit-for-mathematics-in-context-the-pedagogy-of-liberation

Love, B. L. (2019). *We want to do more than survive: Abolitionist teaching and the pursuit of educational freedom.* Beacon Press.

Loveless, T. (2013). *The resurgence of ability grouping and persistence of tracking* (Part II, 2013 Brown Center Report on American Education). The Brookings Institution.

McGee, E. O. (2021). *Black, brown, bruised: How racialized STEM education stifles innovation.* Harvard Education Press.

Menakem, R. (2017). *My grandmother's hands: Racialized trauma and the pathway to mending our hearts and bodies.* Central Recovery Press.

Michaels, S., O'Connor, C., & Resnick, L. B. (2008). Deliberative discourse idealized and realized: Accountable talk in the classroom and in civic life. *Studies in Philosophy and Education*, 27(4), 283–297. http://doi.org/10.1007/s11217-007–9071-1

Mistry, R., Nenadal, L., Hazelbaker, T., Griffin, K., & White, E. (2017). Promoting elementary school-age children's understanding of wealth, poverty, and civic engagement. *PS: Political Science & Politics, 50*(4), 1068–1073. http://doi.org/10.1017/S1049096517001329

Moses, R. P., & Cobb, C. E. (2001). *Radical equations: Civil rights from Mississippi to the Algebra Project*. Beacon Press.

NCSM & TODOS: Mathematics for ALL. (2016). *Mathematics education through the lens of social justice: Acknowledgement, actions, and accountability*. https://www.todos.math/org/socialjustice

National Council of Teachers of Mathematics. (1989). *Curriculum and evaluation standards for school mathematics*. Author.

National Council of Teachers of Mathematics. (2014). *Principles to actions: Ensuring mathematical success for all*. Author.

National Council of Teachers of Mathematics. (2020). *Catalyzing change in early childhood and elementary mathematics: Initiating critical conversations*. Author.

National Education Association. (2015). *Research spotlight on ability grouping*. http://www.nea.org/tools/16899.htm

National Governors Association Center for Best Practices, Council of Chief State School Officers. (2010). *Common Core State Standards: Mathematics*. Author.

National Park Service. (2021). *The Be Straw Free Campaign*. https://bit.ly/3EIYQRK

National Research Council. (2000). *How people learn: Brain, mind, experience, and school*. National Academies Press.

National Research Council. (2001). *Adding it up: Helping children learn mathematics*. National Academies Press.

Nell, M. L., Drew, W. F., & Bush, D. E. (2013). *From play to practice: Connecting teachers' play to children's learning*. National Association for the Education of Young Children.

Nickel City Housing Coop. (n.d.). *Implicit bias discussion questions*. https://bit.ly/335Yuqp

Phillips, J. (2020, May 15). "Why COVID-19 Is a Social Justice Issue," *LATogether*. https://bit.ly/3pF20zz

Pitts, J. (2016). Don't say nothing. *Learning for Justice*. https://www.learningforjustice.org/magazine/fall-2016/dont-say-nothing

Ritchie, H., & Roser, M. (2018, September). Plastic pollution. *Our World in Data*. https://ourworldindata.org/plastic-pollution

Rubel, L., Lim, V., Hall-Wieckert, M., & Sullivan, M. (2016). Teaching mathematics for spatial justice: An investigation of the lottery. *Cognition and Instruction, 34*(1), 1–26. https://doi.org/10.1080/07370008.2015.1118691

School Nutrition Association. (2012). *We can do this: Advice and resources for meeting the NSLP New Meal Pattern*. https://schoolnutrition.org/uploadedFiles/Resources_and_Research/Operations/We%20Can%20Do%20This-Advice%20and%20Resources%20for%20Meeting%20the%20New%20Meal%20Pattern.pdf

Schwartz, D. (2004). *How much is a million?* Lothrop Lee & Shepart Books.

Silver, E. A., & Mills, V. L. (2018). *A fresh look at formative assessment in teaching mathematics*. NCTM.

Singleton, G. E., & Linton, C. (2005). *Courageous conversations about race: A field guide for achieving equity in schools*. Corwin.

Smith, M. S., & Stein, M. K. (2018). *5 practices for orchestrating productive mathematics discussions* (2nd ed.). NCTM.

Steele, D.M., & Cohn-Vargas, B. (2013). *Identity safe classrooms: Places to belong and learn*. Corwin.

Stefanakis, E. (2002). *Multiple intelligences and portfolios*. Heinemann.

Swalwell, K. (2013). *Educating activist allies: Social justice pedagogy with the suburban and urban elite*. Routledge.

United Nations Division for Social Policy. (2006). *Social justice in an open world: The role of the United Nations*. United Nations Publications.

U.S. Department of Housing and Urban Development. (2018). *The 2018 Annual Homeless Assessment Report (AHAR) to Congress, Part 1: Point-in-time estimates of homelessness*. https://www.wpr.org/sites/default/files/2018-ahar-part-1-compressed.pdf

Vasquez, V. M. (2016). *Critical literacy across the K–6 curriculum*. Taylor & Francis.

WWF-Australia. (2019, June 12). *Revealed: plastic ingestion by people could be equating to a credit card a week*. https://bit.ly/3rR67LD

Yeh, C. (under review). *Towards Justice-Oriented Mathematics*.

Yeh, C. & Chao, T. (2019). Celebrating the mathematical brilliance of all children. *Teaching Children Mathematics, 25*(7), 448. https://doi.org/10.5951/teacchilmath.25.7.0448

Yeh, C., Ellis, M. W., & Hurtado, C. K. (2017). *Reimagining the mathematics classroom: Creating and sustaining productive learning environments*. NCTM.

Zinn Education Project. (n.d.). *Delano Grape Strike*. https://bit.ly/32F9r1N

INDEX

**JENNIFER M. BAY-WILLIAMS,
JOHN J. SANGIOVANNI, ROSALBA SERRANO,
SHERRI MARTINIE, JENNIFER SUH**

Because fluency is so much more
than basic facts and algorithms

Grades K–8

**KAREN S. KARP,
BARBARA J. DOUGHERTY,
SARAH B. BUSH**

A schoolwide solution for students'
mathematics success

Elementary, Middle School, High School

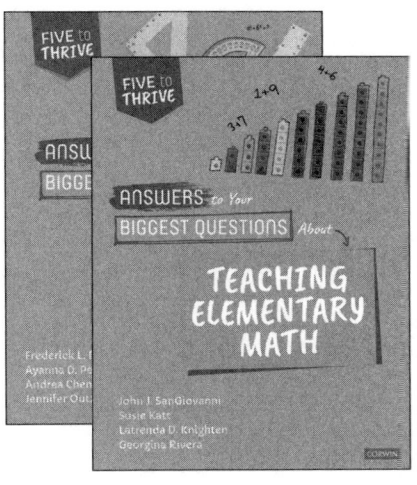

**JOHN J. SANGIOVANNI, SUSIE KATT,
LATRENDA D. KNIGHTEN, GEORGINA RIVERA,
FREDERICK L. DILLON, AYANNA D. PERRY,
ANDREA CHENG, JENNIFER OUTZS**

Actionable answers to your most pressing questions
about teaching elementary and secondary math

Elementary, Secondary

**SARA DELANO MOORE,
KIMBERLY RIMBEY**

A journey toward making
manipulatives meaningful

Grades K–3, 4–8

CORWIN

A SAGE Publishing Company

Helping educators make the greatest impact

CORWIN HAS ONE MISSION: to enhance education through intentional professional learning.

We build long-term relationships with our authors, educators, clients, and associations who partner with us to develop and continuously improve the best evidence-based practices that establish and support lifelong learning.

NATIONAL COUNCIL OF
TEACHERS OF MATHEMATICS

The National Council of Teachers of Mathematics supports and advocates for the highest-quality mathematics teaching and learning for each and every student.